MODERN EARTH SCIENCE

In-Depth
Investigations

HOLT, RINEHART AND WINSTON
Harcourt Brace & Company
Austin • New York • Orlando • Atlanta • San Francisco • Boston • Dallas • Toronto • London

ISBN 0-03-051438-X   12345   021   00   99   98   97

Contents

About In-Depth Investigations

The 30 experiments in *Modern Earth Science In-Depth Investigations* are designed to provide you with practical experience in scientific thought and methods. The investigations reinforce and expand concepts introduced and developed in the textbook. The investigations also give you an opportunity to use scientific methods and to practice laboratory techniques. Each investigation is a process-oriented laboratory—one investigation corresponding to each chapter.

In this booklet, the investigations are four to six pages long. Each investigation has instruction pages and laboratory report pages. The instruction pages provide you with a step-by-step procedure. Each investigation consists of the following sections.

- *Objective* A statement of the purpose of the investigation.
- *Skills* A list of thinking, process, and scientific skills that you will use during the investigation.
- *Introduction* A brief discussion that reviews and expands concepts relevant to the investigation.
- *Materials* A list of materials you will need to conduct the investigation.
- *Prelab Preparation* Lists the references and safety precautions that you should review before beginning the investigation. The prelab preparation may also contain explanations of how to prepare and set up any apparatus you will use during the investigations. There are also questions, calculations, and information that will help you complete the investigation effectively.
- *Procedure* A clear description of the steps you will go through as you conduct the investigation. Throughout the procedure, there are questions that clarify and focus your attention on a specific observation, measurement, or calculation.
- *Analysis and Conclusions* A list of questions that integrate and test your understanding of the concepts presented during the investigation.
- *Extensions* Challenging questions and activities that expand the concepts you mastered during the investigation.

The instruction pages may contain illustrations and sample tables and graphs necessary to clarify concepts or to help organize data. Safety symbols and cautions are included wherever applicable.

The laboratory report data pages consist of a restatement of the objective and a list of the questions presented throughout the prelab preparation, the procedure, the analysis and conclusions, and the extensions. After each question, blank lines or spaces are provided on which you can write your answer or draw a requested illustration. There also may be tables or charts in which you can record data and observations. Some investigations have grids on which you can graph your data.

Laboratory Safety

Many laboratory and field investigations require you to use chemicals and equipment that could cause injury if proper safety guidelines are not used. Advance planning is essential. You should plan carefully for these investigations and be certain that you and your partner are aware of the safety guidelines that must be followed. All the safety guidelines in the investigations are in boldfaced type and begin with the word CAUTION. Be sure to read and follow each caution statement. Safety guidelines that you will find in the investigations are by a safety symbol in the margin; these include:

Electrical Safety

- Never handle electrical equipment with wet hands. Work areas, including floors and tables, should be dry.
- Never overload an electrical circuit.
- Make sure all electrical equipment is properly grounded.
- Keep electrical cords away from areas where someone may trip on them, or where the cords can tip over laboratory equipment.

Fire Safety

- Make sure that fire extinguishers and fire blankets are available in the laboratory.
- Tie back long hair and confine loose clothing.
- Wear safety goggles when working with flames.
- Never reach across an open flame.

Gas Precaution

- Do not inhale fumes directly. When instructed to smell a substance, wave fumes toward your nose and inhale gently.
- Use flammable liquids only in small amounts and in a well-ventilated room or under a fume hood.
- Always use a fume hood when working with toxic or flammable fumes.
- Do not breathe pure gases such as hydrogen, argon, helium, nitrogen, or high concentrations of carbon dioxide.

Glassware Safety

- Check the condition of glassware before and after using it. Inform your teacher about any broken, chipped, or cracked glassware; it should not be used.
- Air-dry glassware; do not dry by toweling. Do not use glassware that is not completely dry.
- Do not pick up broken glass with your bare hands.
- Never force glass tubing into rubber stoppers.
- Never place glassware near edges of your work surface.

Proper Waste Disposal

- Clean up the laboratory after you are finished; dispose of paper toweling, etc.
- Follow your teacher's directions regarding proper procedures for waste disposal, especially for chemical disposal.

Heating Safety

- Use proper procedures when lighting Bunsen burners.
- Turn off hot plates, Bunsen burners, and other open flames when not in use.
- Heat flasks or beakers on a ringstand with a wire gauze between the glass and the flame.
- Store hot liquids in heat-resistant glassware. Heat materials only in heat-resistant glassware.
- Turn off gas valves when not in use.

Chemical Safety

Poison

- Never taste any substance in the laboratory. Do not eat or drink from laboratory glassware.
- Do not eat or drink in the laboratory.
- Properly label all bottles and test tubes containing chemicals.
- Never transfer substances with a mouth pipette; use a suction bulb.

Caustic Substances

- Alert your teacher to any chemical spills.
- Do not let acids and bases touch your skin or clothing. If a substance gets on skin, rinse immediately with cool water and alert your teacher.
- Wear your laboratory apron to protect your clothing.
- Never add water to acids; always add acids to water.
- When shaking or heating a test tube containing chemicals, always point the test tube away from yourself and other students.

Explosion Danger

- Use safety shields or screens if there is a potential danger of an explosion or implosion of apparatus.
- Never use an open flame when working with flammable liquids such as ether or alcohol.
- Follow a water-bath procedure to heat solids. Never risk an explosion by heating rocks or minerals directly.

Hand Safety

Avoiding Injuries

- Always wear gloves when cutting, fire polishing, or bending or using glass tubing.
- Use tongs when heating test tubes. Never hold them in your hand.
- Always allow heated materials, including glassware, to cool before handling them.

Hygienic Care

- Always wash your hands after completing the laboratory investigation.
- Keep your hands away from your face and mouth.

Clothing Protection

- Wear laboratory aprons in the laboratory.
- Confine loose clothing.

Eye Safety

- Wear approved safety goggles in the laboratory.
- Make sure an emergency eye-wash station is available in the laboratory.
- Never look directly at the sun, even for short periods of time. Laboratory goggles will not protect your eyes from the sun.

Water Safety

- When working near water, always work with a partner or adult.
- Always wear a life jacket.
- Do not work near water during stormy weather.

M O D E R N   E A R T H   S C I E N C E

Chapter 1: Introduction to Earth Science
In-Depth Investigation: Scientific Method

Objective
In this investigation, you will use scientific method to predict a change in a small area of the environment.

Skills
observing, inferring, interpreting, predicting

Introduction
Not all scientists think alike; nor do they always agree about various theories. However, all scientists use *scientific method*, part of which includes the skills of observing, inferring, and predicting.

In this investigation, you will apply scientific method as you examine a place where puddles often form after rainstorms. You can study the puddle area even when the ground is dry, but it would be best to observe the area again when it is wet. Since water is the most effective agent of change in our environment, you will be able to make many observations.

Materials
meter stick
hand lens

Prelab Preparation
1. Review Chapter 1, Section 1.2 Paths to Discovery: Scientific Methods, pages 9–13.
2. Find out the difference between a quantitative observation and a qualitative observation.

Procedure
1. Examine the area of the puddle and the surrounding area carefully. On a sheet of paper, write the heading "Observations." Make a *numbered* list of what can be seen, heard, smelled, or felt. Sample observations: "The ground where the puddle forms is lower than the surrounding area," "There are cracks in the soil," and so forth. *Note: Avoid any suggestions of causes.*
2. On another sheet of paper, write the heading "Inferences." Review your observations and write possible causes for those observations. Sample: "Cracks in the soil (Observation 2) may have been caused by lack of rain (Observation 5)."
3. Review your observations and possible causes, and place them into similar groups if possible. Can one cause or set of causes explain several observations? Is each cause reasonable when compared with the others? Does any cause contradict any of the other observations?
4. Start a new page labeled "Hypotheses." You have learned that a hypothesis is a possible explanation to a problem or an occurrence. Create a hypothesis for each group of causes and observations above. Look for the underlying and general mechanisms that produce and link these events. Example: "The clays in the soil shrink as they lose water, thereby causing cracks to form."
5. Based only on your hypotheses, make some predictions about what will happen at the puddle as conditions change. Describe the changes you expect and your reasoning. Sample prediction: "The cracks will grow wider as the puddle dries. Any added water will shrink the cracks."
6. Revisit the puddle several times to see if the changes you observe match your predictions.

Analysis and Conclusions

1. Which of your senses did you use most to make your observations? How could you improve observations using this sense?
2. What could you have used to measure, or put into numbers, many of your observations? Is quantitative observation better than qualitative observation? Explain.
3. Can inferences generally be relied on as true? Explain.
4. If your predictions are found to be incorrect, was the act of forming your inferences a waste of time? Explain.
5. When knowledge is derived from observation and prediction, this process is called "scientific method." After reporting the results of a prediction, how might a scientist continue his or her research?

Extensions

1. Decide whether each of the following statements is an observation (O) or an inference (I). *Note: It does not matter whether the statement happens to be true or false.*
   a. Grass is present inside the puddle.
   b. The grass surrounding the puddle is greener and taller than that inside the puddle.
   c. During a rainstorm, some soil is washed into the puddle.
   d. Water always runs downhill.
   e. Gravity causes the water to run downhill.
   f. The soil that washes out of the puddle will eventually become part of a stream.
   g. Brownish water contains suspended soil particles.
   h. The soil particles are suspended because water is flowing fast.
   i. When the rain stops, the puddle water looks clear.
   j. Mud cracks result from drying the soil.
2. Choose another small area to examine, but look for changes caused by a different factor, such as wind. Follow the steps outlined in the investigation to predict changes that will occur in the area. Use scientific method to design an experiment you can test. Briefly describe your experiment and how you tested it.

M O D E R N   E A R T H   S C I E N C E

Chapter 1: Introduction to Earth Science
In-Depth Investigation: Scientific Method

Objective
In this investigation, you will use scientific method to predict a change in a small area of the environment.

Prelab Preparation
Explain the difference between a quantitative observation and a qualitative observation.

_____

_____

_____

_____

_____

_____

Observations
1. Describe the observation you chose to interpret?

_____

_____

_____

2. Describe your results. Was your prediction correct? Explain. If it was not correct, describe what factors may have affected the results, and how you can improve your observations or revise your predictions.

_____

_____

_____

Analysis and Conclusions
1. Which of your senses did you use most to make your observations? How could you improve observations using this sense?

_____

_____

_____

2. What could you have used to measure, or put into numbers, many of your observations? Is quantitative observation better than qualitative observation? Explain.

_____

_____

_____

3. Can inferences generally be relied on as true? Explain.

_____

_____

_____

4. If your predictions are found to be incorrect, was the act of forming your inferences a waste of time? Explain.

_____

_____

_____

5. After reporting the results of a prediction, how might a scientist continue his or her research?

_____

_____

_____

_____

_____

_____

Extensions

1. Decide whether each statement is an inference (I) or an observation (O).
  a. Grass is present inside the puddle. _____
  b. The grass surrounding the puddle is greener and taller than that inside the puddle. _____
  c. During a rainstorm, some soil is washed into the puddle._____
  d. Water always runs downhill. _____
  e. Gravity causes water to run downhill. _____
  f. The soil that washes out of the puddle will eventually become part of a stream. _____
  g. Brownish water contains suspended soil particles. _____
  h. The soil particles are suspended because water is flowing fast. _____
  i. When the rain stops, the puddle water looks clear. _____
  j. Mud cracks result from drying the soil. _____
2. Briefly describe your experiment and how you tested it.

_____

_____

_____

_____

_____

M O D E R N   E A R T H   S C I E N C E

Chapter 2: The Earth in Space

In-Depth Investigation: Earth-Sun Motion

Objective

In this investigation, you will construct a shadow stick in order to identify how changes in a shadow are related to the earth's rotation. You will also determine how the shadow stick can be used to measure time.

Skills

observing, measuring, inferring

Introduction

During the course of a day, the sun seems to move across the sky. This apparent motion is due to the earth's rotation. In ancient times, one of the earliest devices used by people to study the sun's motion was the shadow stick. The shadow stick is a primitive form of a sundial. Before clocks were invented, sundials were the only means of telling time.

You will create your own shadow stick in this investigation. Using the shadow stick will help you see how shadows change in length and position as the earth rotates.

Materials

magnetic compass	notebook paper	wooden board (9″ x 12″)
metric ruler	masking tape	wooden dowel (about 12″ long and $\frac{1}{4}$″ diameter)
pencil	watch	

Prelab Preparation

1. Review Chapter 2, Section 2.2 Movements of the Earth, pages 29–32

2. Constructing a shadow stick: Using a hand or power drill, make a hole in the board at the location shown in Figure 2.1, just large enough to hold the wooden dowel. **CAUTION: If you are using hand or power tools, it is best to allow your parents or another adult skilled in the tool's operation to help you.** Place the dowel in the hole. See Figure 2.1.

   Place a sheet of three-hole notebook paper on the base of

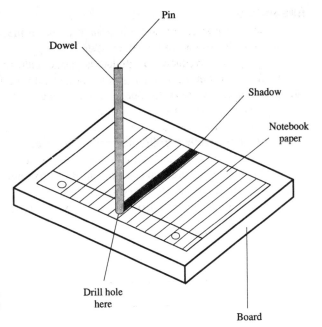

Figure 2.1

the shadow stick. Slip the middle hole of the paper over the stick and slide the paper down so that it rests on the base. Fasten the paper securely to the base with tape. The shadow stick should be just long enough to cast a shadow nearly across the piece of notebook paper.

Procedure

1. Set up the shadow stick in a sunny spot outdoors. **CAUTION: You should never look at the sun.** Align the stick so that its shadow is parallel to the lines of the notebook paper. See Figure 2.1.
2. Place the compass on the paper above the shadow. In one corner of the paper indicate north with an arrow. Label the arrow with a capital N. Why is it necessary to indicate the direction north on the paper?
3. Make a pencil dot in the shadow cast by the wooden dowel. Write the time above the dot. Do this two more times at five-minute intervals. *Note: Do not move the base after you begin to make measurements.*
4. After your last five-minute measurement, wait 10 minutes. Make another dot to show the position of the end of the shadow. Do this two more times at five-minute intervals. Be sure to record the time above each dot. Then return the shadow stick to your classroom.
5. Remove the notebook paper from the base of the shadow stick. Connect the dots with a thin pencil line. Is the line connecting the dots on the paper a straight line?
6. Draw an arrowhead on the end of the line to show the direction in which the shadow moved. In what direction did the shadow move?
7. Make two measurements of the shadow's length in centimeters. The first measurement should be from the center of the paper hole to the first dot. The second measurement should be from the paper hole to the last dot. Record the two lengths.
8. Measure and record the length, in centimeters, of the line connecting the dots. This is the distance the shadow moved in 30 minutes.

Analysis and Conclusions

1. In what direction did the sun appear to move in the 30-minute period?
2. In what direction does the earth rotate?
3. If you made your shadow stick half as long, would its shadow move the same distance in 30 minutes? Explain.
4. How might a shadow stick be used for telling time?

Extension

Repeat this investigation at different hours of the day. Do it early in the morning, early in the afternoon, and early in the evening. Record the results and any differences that you observe. Explain how shadow sticks can be used to tell direction.

M O D E R N   E A R T H   S C I E N C E

Chapter 2: The Earth in Space
In-Depth Investigation: Earth-Sun Motion

Objective
In this investigation, you will construct a shadow stick in order to identify how changes in a shadow are related to the earth's rotation. You will also determine how the shadow stick can be used to measure time.

Observations
1. Why is it necessary to draw a line on the notebook paper showing the direction north?

_____

_____

2. Is the line connecting the dots on the paper a straight line?

_____

_____

3. In what direction did the shadow move?

_____

_____

4. a. What is the length, in centimeters, of the first shadow?

_____

_____

   b. What is the length, in centimeters, of the last shadow?

_____

_____

5. What distance, in centimeters, did the shadow move in 30 minutes?

_____

_____

Analysis and Conclusions
1. In what direction did the sun appear to move in the 30-minute period?

_____

_____

2. In what direction does the earth rotate?

_____

_____

3. If you made your shadow stick half as long, would its shadow move the same distance in 30 minutes? Explain.

_____

_____

4. How might a shadow stick be used for telling time?

_____

_____

_____

_____

_____

Extension

Repeat this investigation at different hours of the day, such as early in the morning, early in the afternoon, and early in the evening. Record the results and any differences that you observe. How can shadows be used to tell whether it is morning or afternoon? How can shadow sticks be used to tell direction?

_____

_____

_____

_____

_____

M O D E R N   E A R T H   S C I E N C E

Chapter 3: Models of the Earth

In-Depth Investigation: Contour Maps—
Island Construction

Objective

In this investigation you will use a contour map to construct a three-dimensional clay model of an island.

Skills

analyzing, comparing, constructing and interpreting models, interpreting data, measuring, reading maps

Introduction

A map is a drawing that shows a simplified version of some detail of the earth's surface. Many different types of maps are available. Each type of map has its own special features and purpose. One of the most basic and useful maps is the contour map. This type of map shows the elevation of the landscape as well as other important features. A contour map is made after a careful survey and photographic study of the area it represents.

Materials

thick wooden dowel or rolling pin	scissors	metric ruler	modeling clay (4 lbs.)
flat basin or large pan (8 cm deep)	pencil	plastic knife	water paper

Prelab Preparation

1. Review Chapter 3, Section 3.3 Topographic Maps, pages 50–55.
2. Study the features of the island contour map in Figure 3.1 on page 17. The elevation is measured in meters.

Procedure

1. Record the contour interval used on the island contour map. Then count the number of contour lines that appear on the map.
2. Roll out as many flat squares of clay as the number of contour lines counted in Step 1. Make each layer about 1 cm thick. Also make the clay layers the same size or larger than the island of your contour map.
3. On a blank sheet of paper, trace the island contour map. Using the scissors, cut out the island from the contour map at the outermost contour line.
4. Place this cutout on top of one of the squares of clay. Trace the edge of the cutout in the clay and remove all clay outside the tracing.
5. Cut out again along the next contour line. Place the paper ring of the contour line that you just cut out on top of the first layer of clay so that the edges line up.
6. Repeat Step 4 with the island map and a new layer of clay.
7. Stack the second layer of clay on the first layer so it fits inside the contour ring you placed on the first layer of clay. This will give you the same contour spacing as shown on the original map. Remove the paper ring.
8. Continue Steps 4–7 for each of the contour layers.
9. Use leftover clay to smooth the terraced edges into a more accurate profile.
10. Make a mark inside the pan approximately 1 cm down from the rim. Put the clay model of the island into the pan and add 1 cm of water. Compare the shoreline with the lines on the contour map. Continue to add water at 1-cm intervals until the water reaches the mark on the pan.

Analysis and Conclusions

1. What is the contour interval of your map?
2. How could you tell the steepest slope from the gentlest slope from observing the spacing of the contour lines?
3. What is the elevation above sea level for the highest point of your model?
4. How do you know if there are any points on your model that are below sea level? If there is such an area, where is this location and what is its elevation?
5. What landscape feature is located at C on your model?
6. What is the elevation of point B on your model?
7. Is there a bench mark? If so, what is its elevation?

Extension

Based on observations of your model, what conclusions can you make about where people would live on this island? Explain.

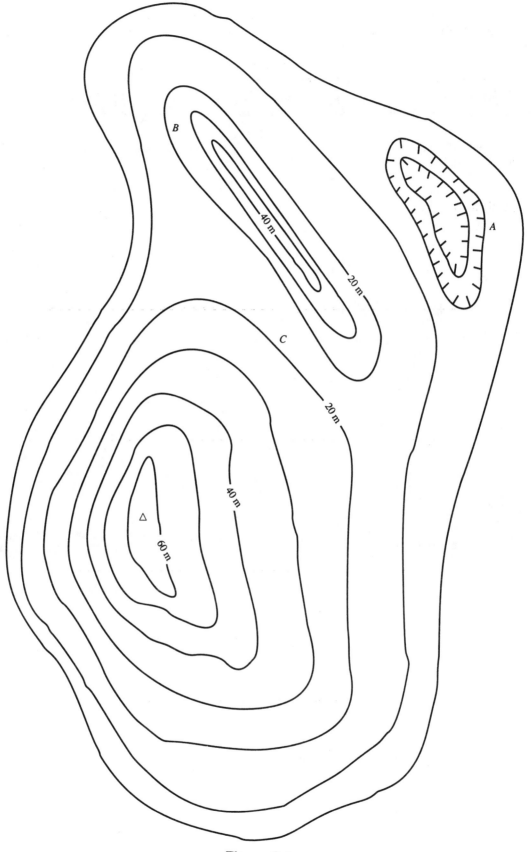

Figure 3.1

M O D E R N   E A R T H   S C I E N C E

Chapter 3: Models of the Earth
In-Depth Investigation: Contour Maps—
Island Construction

Objective
In this investigation you will use a contour map to construct a three-dimensional clay model of an island.

Analysis and Conclusions
1. What is the contour interval of your map?

   _____

2. How could you tell the steepest slope from the gentlest slope from observing the spacing of the contour lines?

   _____

   _____

   _____

   _____

3. What is the elevation above sea level for the highest point of your model?

   _____

   _____

4. How do you know if there are any points on your model that are below sea level? If there is such an area, where is this location and what is its elevation?

   _____

   _____

   _____

5. What feature is located at *C* on your model?

   _____

   _____

6. What is the elevation of point *B* on your model?

   _____

7. Is there a bench mark? If so, what is the elevation?

   _____

Extension
Based on observations of your model, what conclusions can you make about where people would live on this island? Explain.

   _____

   _____

   _____

   _____

Laboratory Notes

M O D E R N E A R T H S C I E N C E

Chapter 4: Plate Tectonics

In-Depth Investigation: A Model of Convection Currents

Objective

In this investigation, you will demonstrate convection-current action and attempt to show how convection currents may be the cause of plate motion.

Skills

observing, measuring, experimenting, predicting, interpreting a model

Introduction

Mid-ocean ridges are places where heat from the mantle reaches the earth's surface. Many geologists think that mid-ocean ridges mark the location of the rising portions of convection currents in the semi-molten asthenosphere. The drag on the lithosphere from such currents might provide the energy for plate motion.

Materials

rectangular aluminum pan
  (at least 25 × 30 × 5 cm)
cold water
beaker (1 liter or larger)
4 ring stands and clamps
lab apron

2 Bunsen burners with flame spreader attachments
2 craft sticks
metric ruler
food coloring with dropper
pencil
3 thermometers

Prelab Preparation

1. Review Chapter 4, Section 4.2 The Theory of Plate Tectonics, pages 72–77.
2. Review the safety guidelines for heating safety and glassware safety.
3. Draw and label a cross-section of a mid-ocean ridge system that lies above a convection current in the asthenosphere.

Procedure

1. Set up the equipment as shown in Figure 4.1. The bottom of the pan should be horizontal and about 6 cm above the top of the flame spreaders of the Bunsen burners.

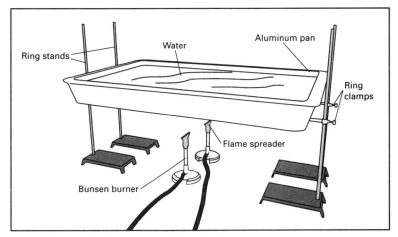

Figure 4.1

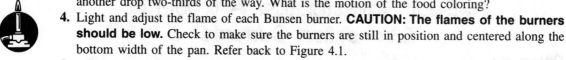

2. Use the beaker or another container to fill the pan with cold water to a depth of at least 4 cm.

3. Imagine a line down the length of the pan that would divide it in half. Place one drop of food coloring into the water about one-third of the way along the imaginary line, and place another drop two-thirds of the way. What is the motion of the food coloring?

4. Light and adjust the flame of each Bunsen burner. **CAUTION: The flames of the burners should be low.** Check to make sure the burners are still in position and centered along the bottom width of the pan. Refer back to Figure 4.1.

5. After a minute or two, tiny bubbles will appear in the water above the Bunsen burners. When this occurs, place another drop of food coloring into each of the positions chosen in Step 3. Describe any differences in the movement of the food coloring compared with earlier movements.

6. After another minute or so, gently place one of the craft sticks onto the water's surface about 3 cm to the left of the center of the pan. Now place the second stick to the right of the center of the pan. Make the strips parallel to each other and parallel to the ends of the pan. Use the pencil to line up the sticks.

7. As soon as you observe the sticks moving, place two more drops of food coloring in the same positions as in Step 3. Describe the motion of the food coloring when the sticks began to move. What is the relation between the motion of the sticks and the motion of the water?

8. With the help of a partner, hold one thermometer bulb just under the water's surface at the center of the pan. Hold the other two thermometers in similar positions near the ends of the pan. Record the temperatures. How does the temperature of the water relate to the motion of the food coloring in the water?

9. Turn off the Bunsen burners. After the pan has completely cooled, carefully empty the water into a sink.

Analysis and Conclusions

1. In relation to plate tectonics, what do the Bunsen burners, water, and craft sticks represent?
2. Explain how the motion of the water affected the motion of the craft sticks.
3. Does the temperature of the water at different places relate to the flow pattern that you observed with the food coloring? Explain.

Extensions

1. Suggest some reasonable modifications of the investigation that would make it a more realistic illustration of the process of seafloor spreading.
2. Find a diagram that shows measurements of heat flow from ocean crust. Do the readings show a pattern of high heat flow at the mid-ocean ridges?

M O D E R N E A R T H S C I E N C E

Chapter 4: Plate Tectonics

In-Depth Investigation: A Model of Convection Currents

Objectives

In this investigation you will demonstrate convection-current action. You will attempt to show how convection currents may be the cause of plate motion.

Prelab Preparation

Make a cross-sectional diagram to show what convection currents under a mid-ocean ridge might look like.

Observations

1. What was the motion of the food coloring in the water before the Bunsen burners were lighted?

   _____

   _____

2. Describe any differences in the movement of the food coloring just after the burners were lighted compared with earlier movements.

   _____

   _____

3. Describe the motion of the food coloring when the sticks began to move.

   _____

   _____

4. What is the relationship between the motion of the sticks and the motion of the water?

   _____

   _____

5. Record the thermometer readings.

Center thermometer _____

End thermometer #1 _____

End thermometer #2 _____

6. Describe how the temperature of the water relates to the motion of the food coloring in the water.

_____

_____

_____

_____

Analysis and Conclusions

1. In relation to plate tectonics, what do the Bunsen burners, water, and craft sticks represent?

_____

_____

2. Explain how the motion of the water affected the motion of the craft sticks.

_____

_____

3. Does the temperature of the water at different places relate to the flow pattern that you observed with the food coloring? Explain.

_____

_____

_____

Extensions

1. Suggest some reasonable modifications of the investigation that would make it a more realistic illustration of the process of seafloor spreading.

_____

_____

_____

2. Find a diagram that shows measurements of heat flow from ocean crust. Do the readings show a pattern of high heat flow at the mid-ocean ridges?

_____

_____

_____

M O D E R N   E A R T H   S C I E N C E

Chapter 5: Deformation of the Crust
In-Depth Investigation: Continental Collisions

Objective
In this investigation you will create a model to help explain how the Himalaya Mountains formed as a result of the collision of India into Eurasia.

Skills
observing, experimenting, constructing and interpreting a model, measuring

Introduction
When the subcontinent of India broke away from Africa and began to move northward toward Eurasia, the oceanic crust on the northern side of India began to subduct beneath the Eurasian plate. The deformation of the crust resulted in mountain formation, and the Himalaya Mountains grew higher and higher. Earthquakes in the Himalayan region suggest that India is still pushing against Eurasia.

Materials
thick cardboard (at least 15 × 25 cm)
adding-machine paper (6 × 35 cm)
paper napkins (light and dark)
metric ruler

wood blocks (2.5 × 2.5 × 6 cm)
5 long bobby pins
masking tape
scissors

Prelab Preparation
1. Review Chapter 4, Section 4.2 The Theory of Plate Tectonics, pages 72–75, and Chapter 5, Section 5.3 Mountain Formation, pages 88–93.
2. Make a cross-sectional diagram to show what happens when a plate carrying oceanic and continental crust collides with a plate carrying continental crust at its edge.

Procedure
1. To assemble the continental-collision model, cut a 7-cm slit in the cardboard about 6 cm from the end of the cardboard and parallel to it. Cut the slit open just wide enough so that the adding-machine paper will feed through it without being loose.
2. Securely tape one wood block lengthwise along the slit between the slit and the near end of the cardboard. Tape the other block to the paper strip lengthwise about 6 cm from one end of the paper. Both blocks should be parallel to one another as shown in Figure 5.1 on page 26.
3. Cut two strips of the light-colored paper napkin about 6 cm wide and 8 cm long. Cut two strips of the dark-colored paper napkin about 6 cm wide and 16 cm long. Fold all four strips in half along their width.
4. Stack the napkin strips on top of each other with all the folds along the same side. Place the two dark-colored napkins on the bottom.
5. Place the napkin strips lengthwise on the paper strip with the ends butted up against the wood block.
6. Attach the napkins to the paper strip using the bobby pins as shown in Figure 5.1.
7. Push the long end of the paper strip through the slit in the cardboard until the first fold of the napkins rests against the other wood block. Hold the cardboard at about eye level and pull gently downward on the paper strip. You may need a partner's help. What happens as the dark-colored paper napkin comes in contact with the fixed block of wood? What happens as you continue to pull downward on the strip of paper? Stop pulling when you feel resistance from the paper strip.

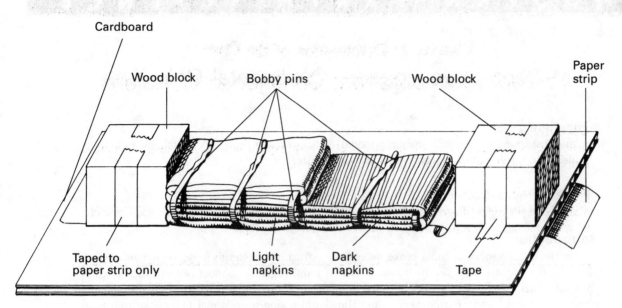

Figure 5.1

Analysis and Conclusions

1. Explain what is represented by the dark napkins, the light napkins, and the wood blocks.
2. What plate tectonics process is represented by the motion of the paper strip in the model? Explain.
3. What type of mountains would result from the kind of collision shown by the model?
4. Explain the differences between the model and the real Himalaya Mountains.

Extensions

1. Obtain a world map of earthquake epicenters. Study the map. Describe the pattern of epicenters in the Himalayan region. Does the pattern suggest that the Himalayas are still growing?
2. Read about the breakup of Gondwanaland and the movement of India toward the northern hemisphere. Write about stages in India's movement. List the time at which each important event occurred.

M O D E R N   E A R T H   S C I E N C E

Chapter 5: Deformation of the Crust

In-Depth Investigation: Continental Collisions

Objectives

In this investigation, you will create a model to help explain how the Himalaya Mountains formed as a result of the collision of India into Eurasia.

Prelab Preparation

Make a cross-sectional diagram to show what happens when a plate carrying oceanic and continental crust collides with a plate carrying continental crust at its edge.

Observations

1. What happens as the dark-colored paper napkin first comes in contact with the stationary block of wood?

2. What happens to the dark-colored and light-colored paper napkins as you continue to pull on the paper strip?

Analysis and Conclusions

1. Explain what is represented by the dark-colored napkins, the light-colored napkins, and the wood blocks.

2. What plate tectonics process is represented by the motion of the paper strip in the model? Explain.

_____

_____

_____

3. What type of mountains would result from the kind of collision shown by the model?

_____

4. Explain differences between the model and the real Himalaya Mountains.

_____

_____

_____

Extensions

1. Study a map of earthquake epicenters. Describe the pattern of epicenters in the Himalayan region. Does the pattern suggest that the Himalayas are still growing?

_____

_____

_____

2. Read about the breakup of Gondwanaland and the movement of India toward the northern hemisphere. Write about stages in India's movement. List the time at which each important event occurred.

_____

_____

_____

_____

_____

_____

Name _____   Class _____   Date _____

M O D E R N   E A R T H   S C I E N C E

Chapter 6: Earthquakes
In-Depth Investigation: Earthquake Waves

Objective
In this investigation, you will find the location of an earthquake's epicenter by applying a method similar to one that scientists use. You will also use a model to demonstrate the difference between two kinds of earthquake waves.

Skills
interpreting data and models, observing, measuring

Part I Introduction
An earthquake releases energy that travels through the earth in all directions. This energy is in the form of waves. Two kinds of earthquake waves are primary waves (P waves) and secondary waves (S waves). Primary waves travel faster than secondary waves and are the first to reach and be recorded at a seismograph station. The secondary waves arrive sometime after the P waves. The time difference between the arrival of the P waves and the S waves increases as the waves travel farther from their origin. This difference in arrival time, called *lag time,* can be used to find the distance to the epicenter of the earthquake. Once the distance from three different locations is determined, scientists can find the exact location of the epicenter.

Materials
drawing compass   ruler   calculator

Prelab Preparation
1. Review Chapter 6, Section 6.2 Recording Earthquakes, pages 103–105.

2. Calculate the following problems, which apply to information in the procedure. The average speed of P waves is 6.1 km/s. The average speed of S waves is 4.1 km/s. To calculate the time it takes seismic waves to travel a given distance, divide that distance by the average speed of each wave.
   a. How long would it take P waves to travel 100 km? How long would it take P waves to travel 200 km?
   b. How long would it take S waves to travel 100 km? How long would it take S waves to travel 200 km?
   c. What is the lag time between the arrival of P waves and S waves over a distance of 100 km? What is the lag time for a distance of 200 km?

Procedure
1. Figure 6.1 on page 30 shows seismograph records made in three cities following an earthquake. The traces begin at the left and arrows indicate the arrival of the P waves. Use the time scale provided to find the lag time between the P waves and the S waves for each city. Be sure to measure the time from the arrival of the P wave to the arrival of the S wave. Record this information in Table 6.1 on your lab report, page 33.

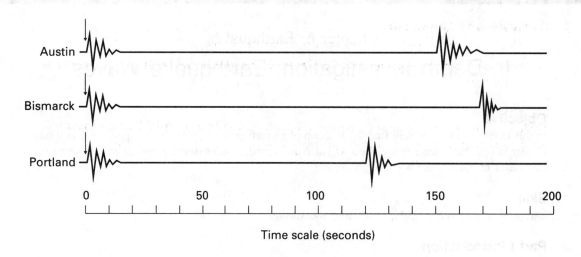

Figure 6.1

2. Find the distance from each city to the epicenter of the earth-quake. To calculate these distances use the lag times you found in Step 1, information from the prelab preparation, and the following formula:

$$\frac{\text{measured lag time (s)} \times 100 \text{ km}}{\text{lag time for 100 km (s)}}$$

Record this information in Table 6.1.

3. Figure 6.2 is a map showing the location of the three cities with a scale in kilometers. Using the map scale on page 34, adjust the compass so that the radius of the circle with Austin at the center is equal to the distance from the epicenter of the earthquake to Austin as calculated in Step 2. Put the point of the compass on Austin. Draw the circle on the map on page 34.

4. Repeat Step 3 for Bismarck and then for Portland. The epicenter of the earthquake is located near the point at which the three circles intersect.

Table 6.1 Epicenter Distances

City	Lag time (seconds)	Distance from city to epicenter
Austin		
Bismarck		
Portland		

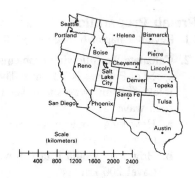

Figure 6.2

Analysis and Conclusion

1. The location of the earthquake epicenter is closest to what city?

2. Why must there be measurements from three different locations to find the epicenter of an earthquake?

Extensions

What is the probability of an earthquake occurring in the area where you live? If an earthquake did occur in your area, what would be its probable cause?

Laboratory Notes

M O D E R N   E A R T H   S C I E N C E

Chapter 6: Earthquakes

In-Depth Investigation: Earthquake Waves

Objective

In this investigation, you will find the location of an earthquake's epicenter by applying a method similar to one that scientists use. You will also use a model to demonstrate the difference between two kinds of earthquake waves.

Prelab Preparation

1. How long would it take P waves moving at 6.1 km/s to travel 100 km? How long would it take P waves to travel 200 km?

_____

2. How long would it take S waves moving at 4.1 km/s to travel 100 km? How long would it take S waves to travel 200 km?

_____

3. What is the time lag between the arrival of P waves and S waves over a distance of 100 km? What is the time lag for a distance of 200 km?

_____

Part I Observations

1. Record your observations in Table 6.1.

Table 6.1 Epicenter Distances

City	Lag time (seconds)	Distance from city to epicenter
Austin		
Bismarck		
Portland		

2. Find the epicenter of the earthquake using the map in Figure 6.2, page 34.

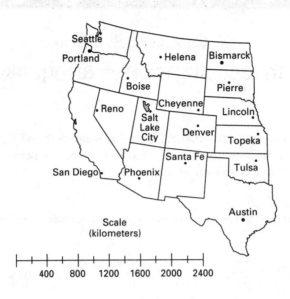

Figure 6.2

Analysis and Conclusions

1. The location of the earthquake epicenter is closest to what city?

_____

2. Why must there be measurements from three different locations to find the epicenter of an earthquake?

_____

_____

_____

Extensions

1. What is the probability of an earthquake occurring in the area where you live?

_____

_____

2. If an earthquake did occur in your area, what would be its probable cause?

_____

_____

_____

M O D E R N   E A R T H   S C I E N C E

Chapter 7: Volcanoes

In-Depth Investigation: Hot Spots and Volcanoes

Objective

In this investigation, you will construct a model to demonstrate how the movement of the Pacific plate is revealed by the orientation of the Hawaiian Islands and associated islands. You will also demonstrate the relationship between hot spots and volcanoes.

Skills

observing, measuring, experimenting, predicting, interpreting a model, reading a map

Introduction

Geologists predict that there are about 100 hot spots on the earth. A hot spot is a place where an extraordinary amount of heat rises from the asthenosphere on a plume of magma. Volcanoes grow where the magma reaches the surface.

The cause of the hot spots is not fully understood. However, it is thought that hot spots maintain their position in the asthenosphere or move very slowly in comparison with plate motion. Therefore, lines of volcanic islands show how the lithospheric plates have moved over the hot spots. By dating the volcanic rocks of such islands, the rate and direction of plate motion are revealed.

Materials

cardboard (about 20 × 30 cm)
white poster board (about 15 × 25 cm)
plastic squeeze bottle (16 mL capacity)
sharp pencil
scissors

red-colored gelatin (about 16 mL)
eyedropper or pipet
metric ruler
lab apron

Prelab Preparation

1. Review Chapter 7, Section 7.1 Volcanoes and Plate Tectonics, pages 117–119; Chapter 4, Section 4.2 The Theory of Plate Tectonics, pages 72–77; Chapter 5, Section 5.3 Mountain Forming, pages 88–93.
2. Obtain a map showing the Pacific plate and its relation to surrounding plates. Study the North Pacific Ocean on the map. Locate the Emperor Seamount chain and the Hawaiian Islands. Count and record the number of individual mountains shown on the map.
3. Mix the gelatin the day prior to the lab.

Procedure

1. Draw a circle about 3 mm in diameter at the center of the cardboard. Using a sharp pencil, make a hole in the cardboard from the underside. *Note: Make the hole slowly and carefully. Do not try to punch or jab at the cardboard with the point.*
2. Again using the sharpened pencil, carefully make six holes in the poster board in the pattern shown in Figure 7.1 on page 36. Each hole should be about 3 mm in diameter. Be sure to make the holes from the underside of the poster board. Write the name ''Pacific Plate'' on the bottom of the paper. Draw an arrow toward the top of the paper, and label it "North." See Figure 7.1.
3. Cut the tip of the squeeze bottle so that the opening is about 3 mm in diameter. Fill the bottle with gelatin using an eyedropper.
4. While your partner holds the cardboard, insert the tip of the squeeze bottle through the hole in the cardboard from underneath.

5. Place the upper left hole in the poster board over the top of the squeeze bottle. Squeeze the bottle so that about 2 mL of gelatin are deposited on the paper around the tip of the bottle. See Figure 7.1.

6. Move the poster board up and left to the next hole, and repeat the procedure. Continue the same procedure for the third and fourth holes.

7. Continue to the fifth and sixth holes, but deposit about 4 mL of gelatin at each hole.

8. Remove the bottle, and place the poster board on your lab table.

9. Write Kure Island, Midway Island, Kauai, and Hawaii next to the appropriate pile of gelatin. Then, label the Emperor Seamounts and the Hawaiian Islands.

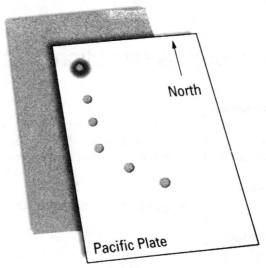

Figure 7.1

10. Write the age "70 million years" next to the first volcano that you made. Label the third volcano "40 million years." Label the last volcano "Presently active."

11. Clean up your work space and store the lab equipment as directed by your teacher.

Analysis and Conclusion

Using your model as a reference, label the appropriate islands on the lines provided in Figure 7.2 on page 37.

1. Describe the direction of motion of the Pacific plate over the last 70 million years as demonstrated by the model.

2. The hot-spot model represents only a small number of volcanoes that exist in the Emperor Seamount chain and the Hawaiian Islands. Based upon the count of volcanoes you completed in the prelab preparation, does it seem that the hot spot now under the Hawaiian Islands has been active for at least 70 million years? Explain.

4. The distance from the southernmost tip of the island of Hawaii to Kure Island is about 2,600 km.

   a. What is the rate of motion of the Pacific plate (in centimeters per year), assuming that the plate traveled that distance in 40 million years?

   b. Geologists estimate that the current rate of motion of the Pacific plate is 2 cm/yr. If this rate

Extension

Obtain a map of Atlantic Ocean seafloor topography to answer the following. Both Iceland and the Azores Islands are located over the Mid-Atlantic Ridge. What are the differences and similarities between these islands and the Hawaiian Island chain?

M O D E R N   E A R T H   S C I E N C E

Chapter 7: Volcanoes

In-Depth Investigation: Hot Spots and Volcanoes

Objective

In this investigation you will construct a model to demonstrate how the movement of the Pacific Plate is revealed by the orientation of the Hawaiian Islands and associated islands. You will also demonstrate the relationship between hot spots and volcanoes.

Prelab Preparation

How many individual volcanoes are shown on the map?

_____

Observations

1. a. What is the relationship between the amount of gelatin deposited on paper and the height of the volcanoes?

_____

b. Does this relationship seem to apply for the real Emperor seamounts and Hawaiian Islands?

_____

2. a. Was the flow of gelatin smooth or was it accompanied by little bursts? What caused the bursts of gelatin? How is this similar to the behavior of magma?

_____

_____

b. How does the behavior of the gelatin compare to the behavior of lava that is building the island of Hawaii?

_____

_____

Analysis and Conclusions

Using your model for reference, label Kure Island, Midway Island, Kauai, and Hawaii on the appropriate lines in Figure 7.2. Then label the Emperor Seamounts and the Hawaiian Islands.

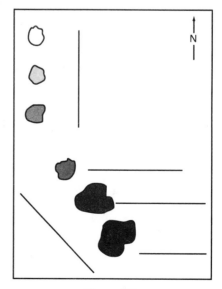

Figure 7.2

1. Describe the direction of motion of the Pacific Plate over the last 70 million years as demonstrated by the model.

   _____

   _____

2. The hot-spot model represents only a small number of volcanoes that exist in the Emperor Seamount chain and the Hawaiian Islands. Based upon the count of volcanoes you completed in the prelab preparation, does it seem that the hot spot now under the Hawaiian Islands has been active for at least 70 million years? Explain.

   _____

   _____

3. The distance from the southernmost tip of the island of Hawaii to Kure Island is about 2,600 km.

   a. What is the rate of motion of the Pacific Plate (in cm/yr), assuming that the plate traveled that distance in 40 million years?

   _____

   b. Geologists estimate that the current rate of motion of the Pacific Plate is 2 cm/yr. If this rate is different from the rate that you determined, what is the explanation for the difference?

   _____

   c. Explain the possible relationship between the rate of plate motion and the size of the volcanic islands in the Hawaiian Islands.

   _____

   _____

Extension

Obtain a map of Atlantic Ocean seafloor topography to answer the following. Both Iceland and the Azores Islands are located over the Mid-Atlantic Ridge. What are the differences and similarities between these islands and the Hawaiian Island chain?

_____

_____

_____

M O D E R N   E A R T H   S C I E N C E

Chapter 8: Earth Chemistry
In-Depth Investigation: Chemical Analysis

Objective
In this investigation, you will use some of the same techniques chemists use to determine the composition of matter. You will test to determine whether water is an element or a compound.

Skills
experimenting, observing, predicting

Introduction
Chemists use a process called *chemical analysis* to learn about the nature and composition of the materials around us. While mixtures can be separated into their component parts by physical means, compounds must be broken down by means of a chemical reaction. Elements cannot be divided by any ordinary chemical means. Therefore, to find out if a substance is a compound or an element, you must carry out a chemical reaction. Chemical reactions involve energy in the form of heat, light, or electricity.

Materials
lab apron	electrodes (stainless steel)	wood splinters
2 connecting wires	beaker (400 mL)	matches
6-volt battery	water	latex gloves
2 test tubes (13 × 100 mm)	Epsom salts (300 g)	balance
stirring rod	safety goggles	

Prelab Preparation
1. Review Chapter 8, Section 8.1 Matter, pages 139–140 and Section 8.2 Combinations of Atoms, pages 147–150.
2. Review general lab safety procedures and proper use of eye and hand protection.
3. Water is a pure substance that is abundant on earth. Do you think water is a compound or an element? Make a hypothesis. How could you find out?

Procedure

1. Set up the apparatus as shown in Figure 8.1. Fill the beaker about three-fourths full with tap water. Connect the battery terminals to the electrodes and observe what happens. What evidence do you see that a reaction is taking place? Is the speed of the reaction the same at each electrode?

2. The rate of the reaction taking place in the beaker is slow. Remove the electrodes from the beaker. *Put on your safety goggles and gloves.* Next, measure out approximately 300 g of Epsom salts and slowly add it to the water in the beaker. Stir the mixture until all or most of the salt has dissolved. The Epsom salts will speed up the reaction because it increases the electrical conductance of the water.

Figure 8.1

3. Fill two test tubes with some of the water–Epsom salt mixture. Disconnect one terminal of the battery, and put the electrodes back in the beaker.

4. Place your gloved index finger over the open end of the first test tube. Now carefully invert the tube and place it below the surface of the mixture in the beaker. Release your finger and note if any air has entered the tube. If so, repeat the inversion process. Follow the same procedure for the second test tube.

5. Place one test tube over each electrode, and reconnect the battery. See Figure 8.2. To speed up the reaction rate, lift the test tubes so that the mouth of each tube just comes to the base of the exposed metal electrode. Use masking tape to hold each tube in this position along the beaker wall. Allow gas to collect until one of the test tubes is full of gas. Disconnect the battery. Which electrode, positive or negative, gave off more gas?

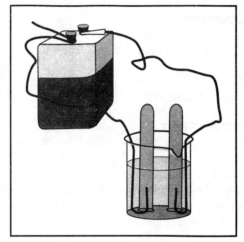

Figure 8.2

6. Test to identify the gases. Hydrogen burns with a colorless flame and makes a popping noise. Oxygen does not burn, but a glowing wood splinter will burst into flames when thrust into oxygen. Test the gas that was collected from the negative electrode first. Slowly remove the test tube from the beaker, keeping it vertical so that any remaining solution can drain into the beaker when it breaks the surface. Once this occurs, quickly place your finger over the mouth of the test tube to keep the gas from escaping. Keep the tube upside down and your finger over the opening until you or your partner brings a lit match to the mouth of the test tube. Be sure to hold the test tube firmly. What happens? What gas is in the test tube?

7. Test the gas in the test tube that was over the positive electrode. Remove the tube and drain any remaining solution as in Step 6, but this time quickly turn the test tube right side up and cover the open end with your finger as before. Uncover the tube, and thrust a glowing splint into the open end. What happens? What gas is in the test tube?

Analysis and Conclusions

1. Is water a compound or an element? Explain.
2. Many tests have been performed to break down hydrogen and oxygen into simpler substances. All these tests have failed. Are hydrogen and oxygen compounds or elements? Explain.

Extensions

1. Baking soda and some antacid tablets fizz when put into water. Is this a physical or a chemical change? How do you know? How might you be able to prove it?
2. When substances such as sugar or salt dissolve in water, is the resulting substance a mixture or a compound? How do you know? How might you be able to prove it?

M O D E R N   E A R T H   S C I E N C E

Chapter 8: Earth Chemistry
In-Depth Investigation: Chemical Analysis

Objective
In this investigation, you will use some of the same techniques chemists use to determine the composition of matter. You will test to determine whether water is an element or a compound.

Prelab Preparation
Do you think water is a compound or an element? How could you find out?

_____

_____

Observations
1. **a.** What evidence do you see that a reaction is taking place?

_____

_____

_____

b. Is the speed of the reaction the same at each electrode?

_____

_____

_____

c. Which electrode, positive or negative, gave off more gas?

_____

_____

2. **a.** What happens when you bring a flame near the mouth of the test tube that was over the negative electrode?

_____

_____

b. What gas is in the test tube?

_____

_____

3. **a.** What happens when you thrust a glowing splinter into the test tube that was over the positive electrode?

_____

_____

_____

b. What gas is in the test tube?

_____

_____

Analysis and Conclusions

1. Is water a compound or an element? Explain.

_____

_____

_____

2. Are hydrogen and oxygen compounds or elements? Explain.

_____

_____

_____

Extensions

1. a. Baking soda and some antacid tablets fizz when put into water. Is this a physical or a chemical change? How do you know?

_____

_____

_____

b. How might you be able to prove it?

_____

_____

_____

2. a. When substances such as sugar or salt dissolve in water, is the resulting substance a mixture or a compound? How do you know?

_____

_____

_____

b. How might you be able to prove it?

_____

_____

_____

M O D E R N E A R T H S C I E N C E

Chapter 9: Minerals of the Earth's Crust
In-Depth Investigation: Mineral Identification

Objective
In this investigation, you will classify several mineral samples using a mineral identification key.

Skills
observing, classifying

Introduction
A mineral identification key can be used to compare the properties of minerals so that unknown mineral samples can be identified. Mineral properties that are often used in mineral identification keys are color, hardness, streak, luster, cleavage, and fracture. Color is probably the most obvious property of minerals. Hardness is determined by a scratch test. The Mohs Scale of Mineral Hardness classifies minerals from 1 (soft) to 10 (hard). Streak is the color of a mineral in a finely powdered form. The streak shows less variation than the color of a sample, and it is more useful in identification. The luster of a mineral is either metallic (having an appearance of metals) or nonmetallic. Some minerals break along defined planes. The planes may be in several directions. This is called cleavage. Other minerals break into irregular fragments. This is called fracture.

Materials
copper penny	sand	streak plate
glass square	steel file	hand lens
mineral samples (5)		

Prelab Preparation
1. Review Chapter 9, Section 9.2 Identifying Minerals, pages 162–169.
2. Most sand grains are composed of the mineral quartz. Look at a sample of quartz sand grains with a hand lens. Are all the sand grains the same color? Is it possible to use only the property of color to identify minerals such as quartz sand?
3. The following is an approximation of the Mohs Hardness Scale:

Can be scratched by	Hardness
Fingernail	2
Copper penny	3
Glass	5
Steel file	6

   What is the hardness of a mineral sample that is scratched by a copper penny but not by a fingernail?

Procedure
1. Study Table 9.1, the Mineral Identification Key, on page 45. Use the table to help you identify the mineral samples. Remember that samples of the same mineral will vary somewhat. Because no specific mineral sample will exactly fit all the properties listed in an identification key, match each sample with the properties that *best* describe it.
2. Perform the observations and tests listed in Table 9.2 on page 44. Observe and record the color of each mineral. Note whether the luster of each mineral is metallic or nonmetallic.

Table 9.2 Mineral Tests

Sample Number	Color/ Luster	Hardness	Streak	Cleavage/ Fracture	Mineral Name
1					
2					
3					
4					
5					

3. Rub each mineral against the streak plate and determine the color of the mineral's streak. Record your observations.

4. Using a fingernail, copper penny, glass square, and a steel file, test each mineral to determine its hardness, based on the hardness scale provided in the prelab activities. Arrange the minerals in order of hardness. Record your observations in Table 9.2.

5. Determine whether the surface of each mineral displays cleavage or fracture. Record your observations.

Analysis and Conclusions

1. Identify your mineral samples. Describe the properties that helped you identify each one.

2. Although color is the most obvious property of a mineral, it is difficult to identify a mineral by its color alone. Explain.

3. a. What is the difference between a scratch test and a streak test?
   b. Why do some very hard minerals leave no streak?

4. Compare the streak to the color of the mineral. Which minerals have the same color as their streak? Which do not?

Extensions

1. Diamonds and graphite are both made of the element carbon, but they are not considered the same mineral. Explain.

2. Corundum, rubies, and sapphires all have the chemical formula Al_2O_3, and they are considered the same types of mineral. Explain.

Table 9.1 Mineral Identification Key

(H = hardness; S = streak)				
Non-metallic light color	Scratches glass	Cleavage	White or pink, two cleavage planes at nearly right angles, H—6, S—white	Orthoclase
		No cleavage	Glassy luster, transparent to opaque, six-sided crystals, H—7, S—white, conchoidal fracture	Quartz
	Does not scratch glass	Cleavage	Colorless to gray, glassy luster three cleavage planes at right angles, H—2.5–3, S—white	Halite
			Colorless to tinted, three cleavage planes not at right angles, double image when you look through it, H—3, S—white	Calcite
			White to pink to colorless, one good cleavage plane, small pieces, flexible, H—1–2.5, S—white	Gypsum
			White to green, soapy feel, one cleavage plane, thin scales, H—1, S—white	Talc
			Colorless to light gray or brown, one cleavage plane, thin sheets, H—2–2.5, S—white	Muscovite
Non-metallic dark color	Scratches glass	Cleavage	Dark green, brown, or black, two cleavage planes at 56° and 124°, H—5–6, S—white	Hornblende
		No cleavage	Tinted red, dull luster, fracture could be taken for cleavage, H—6.5, S—white	Garnet
	Does not scratch glass	Cleavage	Black to dark brown, one cleavage plane, thin sheets, H—2.2–3, S—white to gray	Biotite
		No cleavage	Reddish brown to black, metallic to earthy luster, H—5.5–6.5, S—red to red-brown	Hematite
Metallic luster	Black to dark-green streak		Iron black, some magnetic varieties, H—5–6, S—black	Magnetite
			Black to gray, greasy feel, one cleavage plane, soft, flaky, H—1–2, S—black	Graphite
			Brass yellow, uneven fracture, cubic crystals, H—6–6.5, S—greenish black	Pyrite
			Lead gray, very heavy, three cleavage planes at right angles, H—2.5, S—lead gray to black	Galena

Laboratory Notes

M O D E R N   E A R T H   S C I E N C E

Chapter 9: Minerals of the Earth's Crust
In-Depth Investigation: Mineral Identification

Objective
In this investigation, you will classify several mineral samples using a mineral identification key.

Prelab Preparation
1. Are all sand grains the same color? Is it possible to only use the property of color to identify sand grains?

   _____

2. What is the hardness of a mineral sample that is scratched by a copper penny but not by a fingernail?

   _____

Observations
Record your observations in Table 9.2.

Table 9.2 Mineral Tests

Sample Number	Color/ Luster	Hardness	Streak	Cleavage/ Fracture	Mineral Name
1					
2					
3					
4					
5					

Analysis and Conclusions
1. Identify your mineral samples. Describe the properties that helped you identify each one.

   _____

   _____

   _____

   _____

   _____

2. Although color is the most obvious property of a mineral, it is difficult to identify a mineral by its color alone. Explain.

_____

_____

3. a. What is the difference between a scratch test and a streak test?

_____

_____

_____

b. Why do some very hard minerals leave no streak?

_____

_____

_____

4. Which minerals have the same color as their streak? Which do not?

_____

_____

_____

Extensions

1. Diamonds and graphite are both made of the element carbon, but they are not considered the same mineral. Explain.

_____

_____

_____

2. Corundum, rubies, and sapphires all have the chemical formula Al_2O_3, and they are considered the same types of mineral. Explain.

_____

_____

_____

M O D E R N E A R T H S C I E N C E

Chapter 10: Rocks

In-Depth Investigation: Classification of Rocks

Objective
In this investigation, you will will use a rock identification table to identify various rock samples.

Skills
observing, classifying

Introduction
There are many different types of igneous, sedimentary, and metamorphic rocks. Therefore, it is necessary to know important distinguishing features of the rocks in order to classify them. The classification of rocks is generally based on their mode of origin, their mineral composition, and the size and arrangement (or texture) of their minerals.

The many types of igneous rocks differ in the minerals they contain and the sizes of their crystalline mineral grains. Igneous rocks composed of large mineral grains have a coarse-grained texture. Some igneous rocks have small mineral grains that cannot be seen with the unaided eye. These types of rocks have a fine-grained texture.

Sedimentary rocks are usually made of fragments of other rocks compressed and cemented together. Some sedimentary rocks have a wide range of sediment sizes, while others may have only one size. Other common features of sedimentary rocks include parallel layers, ripple marks, cross-bedding, and the presence of fossils.

Metamorphic rocks often look similar to igneous rocks, but they have bands of color. Metamorphic rocks with a *foliated* texture have minerals arranged in bands. Metamorphic rocks that do not have bands of minerals are *unfoliated*.

Materials
hand lens	rock samples	safety goggles
medicine dropper	10% dilute hydrochloric acid	

Prelab Preparation
1. Study Table 10.1, the Rock Identification Table, on page 51.
2. Review the safety guidelines for eye safety and the handling of caustic substances.

Procedure
1. In Table 10.2 list the numbers of the rock samples you were given by your teacher.
2. Using a hand lens, study the rock samples. Look for characteristics such as the shape, size, and arrangement of the mineral crystals. For each sample, list in Table 10.2 the distinguishing features that you observe.

Table 10.2 Rock Descriptions

Specimen	Description of Properties	Rock Class	Rock Name

3. Refer to the Rock Identification Table on page 51. Compare the properties for each rock sample that you listed in Table 10.2 to the properties listed in the identification table. If you are unable to identify certain rocks, examine these rock samples again.

4. Certain rocks react with acid, indicating they are composed of calcite. If a rock contains calcite, it will bubble. Using a medicine dropper and 10% dilute hydrochloric acid, test various samples for their reactions. **CAUTION: Wear goggles when working with hydrochloric acid.**

5. Complete Table 10.2. Identify the class of rocks—igneous, metamorphic, or sedimentary—that each rock sample belongs to and the name of the rock.

Analysis and Conclusions

1. What properties were most useful in identifying each rock sample?
2. Were there any samples that you found difficult to identify? Explain.
3. Were there any characteristics common to all the rock samples?
4. How can you distinguish between a sedimentary rock and a foliated metamorphic rock when both have observable layering?

Extensions

1. Name properties that distinguish the following pairs of rocks from one another:
   a. granite and limestone   b. obsidian and sandstone
   c. pumice and slate   d. conglomerate and gneiss

2. Collect a variety of rocks in your area. Use the Rock Identification Table in this investigation to see how many you can classify. How many rocks did you collect that were igneous? How many were sedimentary rocks? How many were metamorphic rocks? After you identify the class of each rock, try to name the rock.

Table 10.1 Rock Identification Table

Description	Rock Class	Rock Name
Coarse-grained; mostly light in color—shades of pink, gray, and white are common	Igneous	Granite
Coarse-grained; mostly dark in color; much heavier than granite or diorite	Igneous	Gabbro
Fine-grained; dark in color; often rings like a bell when struck with a hammer	Igneous	Basalt
Light to dark in color; many holes—spongy appearance; light in weight, may float in water	Igneous	Pumice
Light to dark in color; glassy luster—sometimes translucent; conchoidal features	Igneous	Obsidian
Coarse-grained; foliated; layers of different minerals often give a banded appearance	Metamorphic	Gneiss
Coarse-grained; foliated; quartz abundant, commonly contains garnet; flaky minerals	Metamorphic	Schist
Fine-grained; foliated; cleaves into thin flat plates	Metamorphic	Slate
Coarse-grained; nonfoliated; reacts with acid, effervesces	Metamorphic	Marble
Fine-grained; soft and porous; normally white or buff color	Sedimentary	Chalk
Coarse-grained, over 2 mm; rounded pebbles; some sorting—clay and sand can be seen	Sedimentary	Conglomerate
Medium-grained, $\frac{1}{16}$ to 2 mm; mostly quartz fragments—surface feels sandy	Sedimentary	Sandstone
Microscopic grains; clay composition; smooth surface—hardened mud appearance	Sedimentary	Shale
Coarse to medium-grained; well-preserved fossils are common; soft—can be scratched with a knife; occurs in many colors but usually white-gray; reacts with acid	Sedimentary	Crystalline limestone
Coarse to fine-grained; cube-shaped crystals; normally colorless; does not react with acid	Sedimentary	Halite

Laboratory Notes

M O D E R N   E A R T H   S C I E N C E

Chapter 10: Rocks
In-Depth Investigation: Classification of Rocks

Objective
In this investigation, you will will use a rock identification table to identify various rock samples.

Observations
Record your observations in Table 10.2.

Table 10.2 Rock Descriptions

Specimen	Description of Properties	Rock Class	Rock Name

Analysis and Conclusions
1. What properties were most useful in identifying each rock sample?

_____

_____

_____

2. Were there any samples that you found difficult to identify? Explain.

_____

_____

_____

3. Were there any characteristics common to all the rock samples?

_____

_____

4. How can you distinguish between a sedimentary rock and a foliated metamorphic rock when both have observable layering?

_____

_____

_____

Extensions

1. Name properties that distinguish the following pairs of rocks from one another.
a. granite and limestone:

_____

_____

b. obsidian and sandstone:

_____

_____

c. pumice and slate:

_____

_____

d. conglomerate and gneiss:

_____

_____

2. How many rocks did you collect that were igneous? How many were sedimentary rocks? How many were metamorphic rocks? Try to name the types of rocks you collected.

_____

_____

M O D E R N   E A R T H   S C I E N C E

Chapter 11: Resources and Energy
In-Depth Investigation: Extraction of Copper from Its Ore

Objective
In this investigation, you will extract copper from copper carbonate in much the same way that copper is extracted from malachite ore.

Skills
experimenting, observing, predicting, demonstrating, inferring, interpreting models

Introduction
Most metals are found in the earth's crust combined with other elements. A material in the crust that is a profitable source of metal is called an *ore*. Malachite, a native copper ore, is the basic carbonate of copper, which can be represented by the formula $CuCO_3Cu(OH)_2$. The green corrosion that forms on copper due to weathering has the same composition as malachite. The reactions of malachite ore are similar to those of copper carbonate, $CuCO_3$.

Materials
2 test tubes, 13 × 100 mm	copper (cupric) carbonate	iron filings
funnel	Bunsen burner	test tube rack
water	dilute sulfuric acid	test tube clamp
safety goggles	lab apron	

Prelab Preparation
1. Review Chapter 9, Section 9.1 What Is a Mineral?, pages 157–162; and Chapter 11, Section 11.1 Mineral Resources, pages 195–197.
2. Review the safety guidelines for fire, heating, and chemical safety, and for handling caustic substances.

Procedure
1. **CAUTION: Wear your laboratory apron and safety goggles throughout the investigation.** Fill one of the test tubes about one-fourth full of copper carbonate. Record the color of the copper carbonate.
2. Light the Bunsen burner and adjust the flame.
3. Heat the copper carbonate by holding the tube over the flame with a test-tube holder as shown in Figure 11.1. **CAUTION: When heating a test tube, point it away from yourself and other students. CAUTION: Gently move the test tube over the flame to heat it slowly and prevent the test tube from breaking.** As you heat the copper carbonate, observe any changes in color.

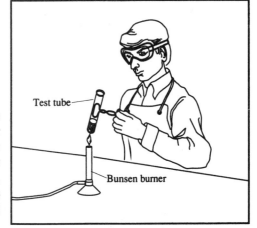

Figure 11.1

4. Continue heating the test tube over the Bunsen burner flame for five minutes. The following chemical reaction is occurring in the test tube:

$$CuCO_3 \rightarrow CuO + CO_2$$

5. Allow the test tube to cool. Then place the test tube in the test tube rack. Insert a funnel in the test tube and add dilute sulfuric acid until the test tube is 3/4 full. **CAUTION: Avoid touching the sides of the test tube; it may get hot. If any of the acid gets on your skin or clothing, rinse immediately with cool water and alert your teacher.**

6. Allow the test tube to stand until some of the substance at the bottom of the test tube dissolves. After the sulfuric acid has dissolved some of the solid substance, note the color of the solution.

7. Use the funnel to add more sulfuric acid to the test tube until it is nearly full. Allow the test tube to stand until more of the substance at the bottom of the test tube dissolves. Pour this solution (copper sulfate) into the second test tube.

8. Add a small amount of iron filings to the second test tube. Observe what happens.

9. Clean all laboratory equipment and dispose of the sulfuric acid as directed by your teacher.

Analysis and Conclusions

1. **a.** Disregarding any condensed water on the test-tube walls, what is the new substance formed in the first test tube called?

   b. Does the new substance take up as much space in the test tube as did the copper carbonate? Explain.

2. **a.** When the iron filings were added to the second test tube, what indicated that a chemical reaction was taking place?

   b. Describe any change to the iron filings.

   c. Describe any change in the solution.

3. In the actual process of extracting copper from its ore, the copper sulfate solution is allowed to flow over cast iron scrap metal. The loose layer of copper that forms on the scrap metal is then separated and pressed into bars or redissolved for purification. What do the iron filings represent in the actual process of extracting copper from its ore?

Extension

Suppose that a certain deposit of copper ore contains a minimum of 1.0% copper by mass and that copper sells for 30 cents per kilogram. Approximately how much could you spend to mine and process the copper out of 100 kg of copper ore and remain profitable?

HRW material copyrighted under notice appearing earlier in this work.

57

Laboratory Notes

M O D E R N   E A R T H   S C I E N C E

Chapter 11: Resources and Energy

In-Depth Investigation: Extraction of Copper from Its Ore

Objective

In this investigation, you will extract copper from copper carbonate in much the same way that copper is extracted from malachite ore. You will also demonstrate a method of copper purification.

Prelab Preparation

1. Define the term *electrolysis* as it applies to chemistry.

_____

_____

2. Define the term *weathering*.

_____

Part I Observations

1. What color is copper carbonate?

_____

2. After you heat the copper carbonate, what color is the substance in the test tube?

_____

3. a. What is the new compound formed in the bottom of the test tube?

_____

    b. Does the new compound take up as much room in the test tube as did the original copper carbonate? Explain.

_____

_____

4. What is the color of the liquid solution formed when dilute sulfuric acid is added to the test tube?

_____

5. a. What indicates that a chemical reaction is taking place?

_____

    b. If there is a change in the iron filings describe the change.

_____

    c. If there is a change in the color of the solution, describe the change.

_____

6. In your demonstration, what do the iron filings represent in the actual process of extraction of copper from its ore?

_____

_____

HRW material copyrighted under notice appearing earlier in this work.

59

Analysis and Conclusions

1. a. Disregarding any condensed water on the test-tube walls, what is the new substance formed in the first test tube called?

_____

_____

_____

b. Does the new substance take up as much space in the test tube as did the copper carbonate? Explain.

_____

_____

_____

_____

2. a. When the iron filings were added to the second test tube, what indicated that a chemical reaction was taking place?

_____

_____

_____

b. Describe any change to the iron filings.

_____

_____

_____

c. Describe any change in the solution.

_____

_____

_____

3. In the actual process of extracting copper from its ore, the copper sulfate solution is allowed to flow over cast iron scrap metal. The loose layer of copper that forms on the scrap metal is then separated and pressed into bars or redissolved for purification. What do the iron filings represent in the actual process of extracting copper from its ore?

_____

_____

_____

Extension

Suppose that a certain deposit of copper ore contains a minimum of 1.0% copper by mass and that copper sells for 30 cents per kilogram. Approximately how much could you spend to mine and process the copper out of 100 kg of copper ore and remain profitable?

_____

M O D E R N   E A R T H   S C I E N C E

Chapter 12: Weathering and Erosion
In-Depth Investigation: Soil Chemistry

Objective
In this investigation, you will classify a soil sample as pedocal or pedalfer.

Skills
classifying, experimenting, inferring

Introduction
The kinds and amounts of minerals present in soil are important to life. To support plant life, the soil must have a proper balance of minerals. In order for plants to take in the minerals they need, the soil must also have the proper acidity. Plants use up some minerals, but allow others to accumulate. The minerals that are used up are replaced by the decay of dead plants or by the addition of fertilizers. The minerals that are most likely to be used up contain the elements nitrogen, phosphorus, and potassium. Minerals that tend to accumulate contain aluminum silicates, iron silicates, and calcium carbonate. Soils with an accumulation of aluminum and iron silicates are called *pedalfer* soils. The term "pedalfer" is derived from a combination of the Greek word **ped**on meaning "soil," **al**uminum, and the Latin word **fer**rum meaning "iron." Soils rich in calcium carbonate are called *pedocal* (**ped**on + **calc**ium) soils. These substances accumulate mostly in the B horizon of the soil. Hydrochloric acid has little or no effect on silicates, but it decomposes calcium carbonate, causing CO_2 gas to bubble out of solution.

Materials
ammonia solution	medicine dropper	subsoil sample (B/C horizons)
9 cork stoppers	pH paper	9 test tubes
water	topsoil sample (A horizon)	test-tube rack
dilute hydrochloric acid	safety goggles	lab apron

Prelab Preparation
1. Review Chapter 12, Section 12.3 Weathering and Soil, pages 227–230.
2. Review the safety guidelines for handling caustic substances.
3. Define the terms acidic, alkaline, and neutral.
4. Acidity is measured on a scale called the pH scale. The pH scale ranges from 0 (acidic) to 14 (alkaline). A pH of 7 is neutral (neither acidic nor alkaline). The following is a list of different substances and their pHs. Classify the substances as acidic, alkaline, or neutral.

soft drink	3.0	sea water	8.5	milk	7.0
pure water	7.0	lemon juice	2.2	vinegar	2.8
milk of magnesia	10.5	ammonia	11.5	blood	7.5
orange juice	3.5				

5. The pH paper changes color in the presence of an acid or alkaline substance. Take a strip of pH paper and wet it with tap water. Compare the color of the wet pH paper with the pH color scale. What is the pH of tap water?

Procedure
1. **CAUTION: Wear your laboratory apron and safety goggles.** Place a small amount of the topsoil sample in a clean test tube until 1/8 full. Add water to the test tube until 3/4 full. Stopper the test tube and shake. Set the soil and water mixture aside in the test-tube rack to settle. When the water is fairly clear, test it with a piece of pH paper. What is the pH of the soil sample? Is the soil acidic or alkaline?

HRW material copyrighted under notice appearing earlier in this work.

61

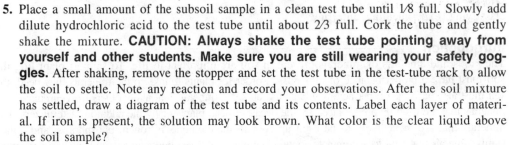

2. Repeat Step 1 with the subsoil sample. What is the pH of this sample? Is the soil acidic or alkaline?

3. Pedalfer soils have a tendency to become acidic. Pedocal soils tend to become alkaline. Based on the pH results in Steps 1 and 2, predict whether your soil is pedalfer or pedocal.

4. To test your prediction, you will need to test the composition of your soil sample. Take out five rock particles from the subsoil sample. Place each particle in a separate test tube. Using the medicine dropper, add two drops of hydrochloric acid to the tubes. **CAUTION: If any of the acid gets on your skin or clothing, rinse immediately with cool water and alert your teacher.** Record your observations. How many of the rock particles were silicates? How many were calcium carbonate?

5. Place a small amount of the subsoil sample in a clean test tube until 1/8 full. Slowly add dilute hydrochloric acid to the test tube until about 2/3 full. Cork the tube and gently shake the mixture. **CAUTION: Always shake the test tube pointing away from yourself and other students. Make sure you are still wearing your safety goggles.** After shaking, remove the stopper and set the test tube in the test-tube rack to allow the soil to settle. Note any reaction and record your observations. After the soil mixture has settled, draw a diagram of the test tube and its contents. Label each layer of material. If iron is present, the solution may look brown. What color is the clear liquid above the soil sample?

6. Using a medicine dropper, place 10 drops of the clear liquid in a clean test tube. Following the same precautions as in Step 4, add 12 drops of ammonia to the test tube. Test the pH of the resulting solution. If the pH is greater than 8, any iron present should settle out as a reddish-brown residue. The remaining solution will be colorless. If the pH is less than 8, add two more drops of the ammonia and test the pH again. Continue until the pH reaches 8 or higher. Record your observations and draw a diagram of the contents of the test tube. Label each layer of material. Is iron present in your soil sample?

Analysis and Conclusions

1. Is your soil sample most likely pedalfer or pedocal? Explain your answer based on the results of the tests performed in Steps 4–6 of the Procedure.

2. **a.** What type of soil, pedalfer or pedocal, would you treat with acidic substances such as phosphoric acid, sulfur, or ammonium sulfate to help plant growth? Explain why.

   b. Explain why the acidic substances in the item above are usually spread on the surface of the soil.

Extension

Why has the use of phosphate and nitrate detergents been banned in some areas?

M O D E R N   E A R T H   S C I E N C E

Chapter 12: Weathering and Erosion
In-Depth Investigation: Soil Chemistry

Objective
In this investigation, you will identify a soil sample as being pedocal or pedalfer.

Prelab Preparation
1. Define the following terms.

   a. acidic _____

   _____

   b. alkaline _____

   _____

   c. neutral _____

2. Refer to page 61 to find the pH for each of the following substances. Then, classify the following substances as acid, alkaline, or neutral.

   a. soft drink _____   **f.** lemon juice _____

   b. pure water _____   **g.** ammonia _____

   c. milk of magnesia _____   **h.** milk _____

   d. orange juice _____   **i.** vinegar _____

   e. sea water _____   **j.** blood _____

3. What is the pH of the tap water?

   _____

Observations
1. **a.** What is the pH of the topsoil sample?

   _____

   b. Is this acid or alkaline?

   _____

2. **a.** What is the pH of the subsoil sample?

   _____

   b. Is this acid or alkaline?

   _____

3. Based on your pH results, hypothesize about whether your soil sample is pedalfer or pedocal.

   _____

4. **a.** How many of the rock particles were silicates?

   _____

   b. How many were calcium carbonate?

   _____

HRW material copyrighted under notice appearing earlier in this work.

63

5. What color is the liquid above the soil sample? Draw the contents of your test tube in test tube *a* shown below. Label each layer of material.

_____

6. Is iron present in your soil sample? Draw the contents of your test tube in test tube *b* shown below.

_____

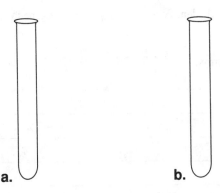

a.   b.

Analysis and Conclusions

1. Is your soil sample pedalfer or pedocal? Explain your answer based on the results of the tests performed in Steps 4–6 of the procedure.

_____

_____

2. a. What type of soil, pedalfer or pedocal, would you treat with acidic substances to help plant growth? Explain why.

_____

_____

   b. Explain why the acidic substances in the item above are usually spread on the surface of the soil.

_____

_____

Extension

Why has the use of phosphate and nitrate detergents been banned in some areas?

_____

_____

_____

MODERN EARTH SCIENCE

Chapter 13: Water and Erosion
In-Depth Investigation: Sediments and Water

Objective
In this investigation you will determine the erosional effect of water on different types of sediment.

Skills
observing, measuring, calculating, experimenting, predicting, interpreting a model, inferring

Introduction
Rain falling on a barren patch of sand or soil causes the surface to become wet. The rainwater then sinks deeper into the sand or soil. The amount of rainwater that can be held by any sediment depends on how much moisture is already in the sediment and the total amount of moisture the sediment can hold. If the water can move through the soil quickly enough, then the soil may never become filled with water. If a sediment is holding all the water it possibly can, then any additional water added to the sediment will cause the water to make puddles or flow downhill. Surface water flow, or runoff, is the primary cause of erosion of barren soil.

Materials
metric ruler	water	large nail
graduated cylinder (100 mL)	grease pencil	timer
dry, coarse sand	two cardboard juice	pan (30 × 25 × 5 cm
dry, silty (unsorted) soil	containers (12-oz.)	or larger)

Prelab Preparation
1. Review Chapter 13, Section 13.2 River Systems, pages 247–251.
2. Review the safety guidelines for hygienic care.
3. Pinch a few drops of water between your thumb and index finger. Hold your fingers at eye level and observe the water as you slowly move your thumb and finger apart. Describe and explain your observations.
4. Wet one of your index fingers. Place both of your index fingers on a lab table or desk and push your fingers away from you along the table's surface. Does the wet finger or the dry finger slide more easily? Explain why.

Procedure
1. Using the graduated cylinder, pour 300 mL of water into each of the juice containers.
2. Place the containers on a flat surface. With the grease pencil, draw a line around the inside of the containers marking the height of the water. On the outside of the containers, label one container A, and the other B. Empty and dry the containers.
3. Fill container A with the silty soil up to the line drawn inside the container. Tap the container gently to even out the surface of the sediment. Add more sediment if needed. Repeat this step, filling container B with sand.
4. Fill the graduated cylinder with 100 mL of water. Slowly pour the water into container A. Stop about every five seconds, allowing the water to be absorbed. Continue pouring until no more water is absorbed. There should be a thin film of water on the surface of the sediment. If more than 100 mL of water is needed, refill the graduated cylinder and continue this step.
5. Record the volume of water that you poured into the container.

6. Repeat Steps 4 and 5 using container B. Which type of sediment held more water?

7. Use the ruler to measure 1 cm above the surface of the sediments in both containers. Draw a line with the grease pencil to mark this height on the inside of both containers. Pour water from the graduated cylinder into container A until it reaches the 1-cm mark.

8. At the very bottom of the side of container A, push in a nail. Place the container inside the pan. At the same time, start the timer and pull the nail out of the container.

9. Observe the water level and record the amount of time it takes the water to drop to the sediment surface.

10. Repeat Steps 8 and 9 using container B. Which sediment was the water able to flow through faster?

11. Empty, clean, and dry the pan and the juice containers. Fill one container with fresh, dry sand and the other container with fresh, dry soil.

12. From a height of 30 cm, empty the container of sand into the right side of the pan, forming a mound. From the same height, empty the container of soil into the left side. Measure and record the approximate heights of each mound in Table 13.1. Which mound is higher?

Table 13.1

Sediment	Mound height (cm)
Dry sand	
Dry soil	
Wet sand	
Wet soil	

13. Slowly add water to each of the mounds until all the sediment is damp. Pack the sediments, making each mound as high as possible without toppling. Wash your hands. Measure and record the heights of the mounds in Table 13.1. Which mound of wet sediment is higher? Is either mound higher than it was when dry? Explain why.

14. Continue adding water to the mounds. What eventually happens? Explain.

Analysis and Conclusions

1. Based on your answer to the questions in Steps 6 and 10, would water erode an area of silty soil or an area of sand more quickly? Explain.

2. Would a hillside covered with sand or a hillside covered with silty soil be more easily eroded during moderate rainfall? Explain with reference to Step 13.

3. Would a hillside covered with sand resist erosion during an extended period of heavy rain? Would a hillside covered with silt resist the erosion?

Extension

Describe ways in which slopes covered with soil can be made more resistant to erosion.

M O D E R N   E A R T H   S C I E N C E

Chapter 13: Water and Erosion

In-Depth Investigation: Sediments and Water

Objective
In this investigation you will determine the erosional effect of water on different types of sediment.

Prelab Preparation
1. Describe and explain your observations of the water you pinched between your thumb and index finger.

   _____

   _____

   _____

   _____

2. Does your wet finger or dry finger slide more easily along the table's surface? Explain why.

   _____

Observations
1. What was the volume of water poured into container A? What was the volume poured into container B?

   Container A _____

   Container B _____

2. Which sediment held more water?

   _____

3. How long did it take for the water level to drop to the surface of the sand? How long did it take the water level to drop to the surface of the silty soil?

   Sand _____

   Silty soil _____

4. Through which sediment was the water able to flow faster?

   _____

5. Record your measurements of mound heights in Table 13.1.

Table 13.1

Sediment	Mound height (cm)
Dry sand	
Dry soil	
Wet sand	
Wet soil	

6. Which mound of dry sediment is higher, the soil or the sand?

_____

7. Which mound of wet sediment is higher?

_____

8. Is either wet mound higher than it was when dry? Explain why.

_____

_____

9. What happens as more water is added to the sediments? Explain why.

_____

_____

_____

Analysis and Conclusions

1. Based on your answers to Observations 2 and 4, would water erode an area of silty soil or an area of sand more quickly? Explain.

_____

_____

_____

2. Would a hillside covered with sand or a hillside covered with silty soil be more easily eroded during moderate rainfall? Explain with reference to Observations 6 and 7.

_____

_____

_____

3. Would a hillside covered with sand resist erosion during an extended period of heavy rain? Would a hillside covered with silt resist the erosion? Relate your answer to Observation 9.

_____

_____

_____

Extension

Describe ways in which slopes covered with soil can be made more resistant to erosion.

_____

_____

_____

M O D E R N   E A R T H   S C I E N C E

Chapter 14: Groundwater and Erosion
In-Depth Investigation: Porosity

Objective
In this investigation, you will measure and compare the porosity of three samples that represent rock particles.

Skills
measuring, observing, interpreting a model

Introduction
Whether soil is composed of coarse pieces of rock or very fine particles, there is always some empty space between the pieces of solid material. This empty space is called pore space. Porosity is calculated by dividing the volume of the pore space by the total volume of the soil sample. Thus, if 50 cm^3 of soil contains 5.0 cm^3 of pore space, then the porosity of the soil sample is 5.0 cm^3/50 cm^3 = 0.10 × 100 = 10%. The result is generally written as a percentage.

Materials
beaker (100 mL)   plastic beads (4 mm)   plastic beads (8 mm)
graduated cylinder (100 mL)

Prelab Preparation
1. Review Chapter 14, Section 14.1 Water Beneath the Surface, pages 261–262.
2. What is a *meniscus*?

Procedure
1. Use a graduated cylinder and water to determine the volume of the beaker. Record the volume.
2. Dry the beaker and fill it to the top with large (8 mm) plastic beads. Gently tap the beaker to settle and compact the beads. Add more beads as needed to fill the beaker until the beads are level with the top. Do the beads represent well-sorted large rock particles, well-sorted small rock particles, or unsorted rock particles? What is the total volume of the beads, including the pore space?
3. Fill the graduated cylinder with water to the top mark and note the position of the water level. Carefully pour the water from the cylinder into the beaker with the large beads until the water level just reaches the top of the beads. Check the new water level in the graduated cylinder. How much water did you add? What is the volume of the pore space?
4. Calculate the porosity of the beads. Record the porosity as a decimal and as a percentage.
5. Repeat Steps 2–4 using the small (4 mm) plastic beads. Do the beads represent well-sorted large rock particles, well-sorted small rock particles, or unsorted rock particles? What is the volume of the pore space between the small plastic beads? What is the porosity of the small beads?
6. Drain and dry both sets of beads. Mix together equal volumes of the small and large beads. Repeat Steps 2–4 using the mixed-size beads. Do the beads represent well-sorted or unsorted rock particles?

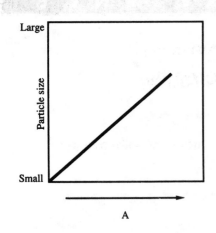
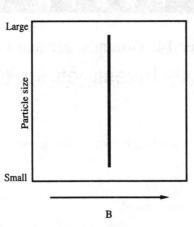
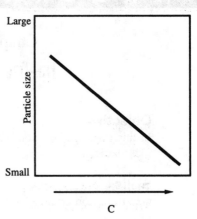

Figure 14.1

Analysis and Conclusions

1. Compare the porosity of the large beads with the porosity of the small beads.
2. Does porosity depend upon particle size? Explain.
3. What effect did mixing the bead sizes have on the porosity? Explain this effect.

Extensions

1. Each of the three graphs in Figure 14.1 represents one of the following properties plotted against particle size: porosity, permeability, and capillarity. Particle size is plotted on the vertical axis, increasing from bottom to top. Porosity, permeability, and capillarity are plotted on the horizontal axis, increasing from left to right.
   a. Which graph represents porosity? Explain.
   b. Which graph represents permeability? Explain.
   c. Which graph represents capillarity? Explain.
2. What would be the effect on the porosity of coarse gravel if it were mixed with fine sand? Conduct an experiment to find out if your answer is accurate.

M O D E R N   E A R T H   S C I E N C E

Chapter 14: Groundwater and Erosion
In-Depth Investigation: Porosity

Objective
In this investigation, you will measure and compare the porosity of three samples that represent rock particles.

Prelab Preparation
What is a *meniscus*?

_____

_____

_____

Observations
1. What is the volume of the beaker?

_____

2. Do the 7-mm beads represent well-sorted large rock particles, well-sorted small rock particles, or unsorted rock particles?

_____

3. What is the total volume of the large beads, including the pore space?

_____

4. How much water did you add to the beaker?

_____

_____

5. What is the volume of the pore space between the large beads?

_____

6. What is the porosity of the large plastic beads? (Give your answer as a decimal and also as a percent.)

_____

7. Do the 4-mm plastic beads represent well-sorted large rock particles, well-sorted small rock particles, or unsorted rock particles?

_____

8. a. What is the volume of the pore space between the small plastic beads?

_____

   b. What is the porosity of the small beads?

_____

9. Do the equal volumes of the small and large beads represent well-sorted or unsorted rock particles?

_____

_____

Analysis and Conclusions

1. Compare the large-bead porosity with the small-bead porosity.

_____

2. Does porosity depend on particle size? Explain.

_____

_____

_____

_____

3. What effect did mixing the bead sizes have on the porosity? Explain this effect.

_____

_____

Extensions

1. Each of the three graphs in Figure 14.1 represents one of the following properties plotted against particle size: porosity, permeability, and capillarity. Particle size is plotted on the vertical axis, increasing from bottom to top. Porosity, permeability, and capillarity are plotted on the horizontal axis, increasing from left to right.

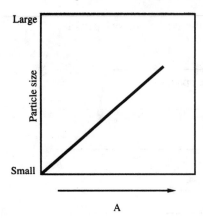
A

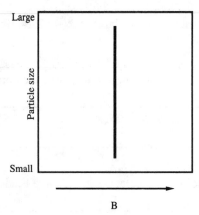
B

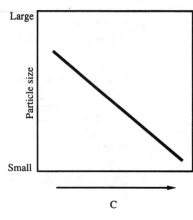
C

Figure 14.1

    a. Which graph represents porosity? Explain.

_____

    b. Which graph represents permeability? Explain.

_____

    c. Which graph represents capillarity? Explain.

_____

2. What would be the effect on the porosity of coarse gravel if it were mixed with fine sand? Describe the experiment you conducted to find the answer.

_____

_____

_____

M O D E R N   E A R T H   S C I E N C E

Chapter 15: Glaciers and Erosion
In-Depth Investigation: Glaciers and Sea Level

Objective
In this investigation, you will construct a model to simulate what would happen if the Antarctic ice sheet melted.

Skills
constructing and interpreting a model, observing, inferring

Introduction
Today, glaciers hold only about 2.2% of the earth's water, but if the polar ice sheets melted, the coastal areas of many countries would be flooded. In the United States, many major cities, such as New York, New Orleans, Houston, and Los Angeles would flood if the sea level rose only a few meters.

Materials
metric ruler	sand and small pebbles (3–5 lb.)	water
milk carton (small)	shallow pan (30 × 25 × 5 cm)	wooden block (5 × 5 × 5 cm)

Prelab Preparation
1. Review Chapter 15, Section 15.1 Glaciers: Moving Ice, pages 277–281.
2. The day before the investigation, fill the milk carton with water and place it in a freezer.
3. Area (A) is calculated by multiplying length (l) times width (W) [$a = l \times w$]. Area is expressed in square units. Volume (V) is calculated by multiplying length (l) times width (w) times height (h). Volume is expressed in cubic units. Calculate the area of the two-dimensional shape and the volume of the three-dimensional shape shown in Figure 15.1.

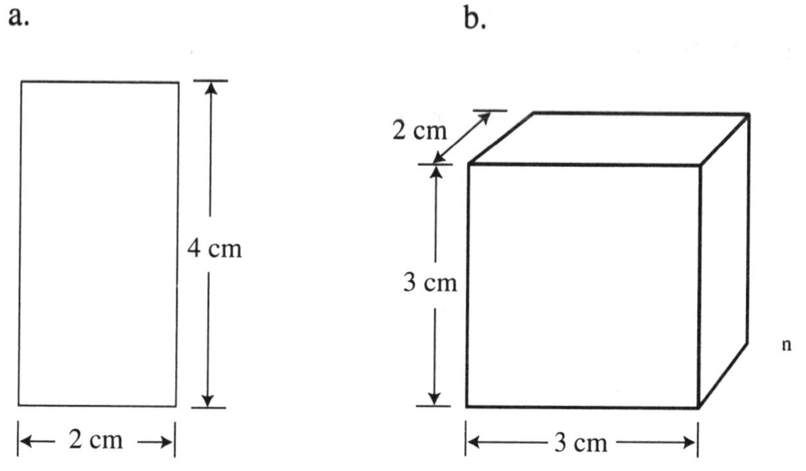

a.

b.

4 cm

2 cm

2 cm

3 cm

3 cm

n

Figure 15.1

Procedure
1. Calculate and record the approximate surface area of the bottom of the pan.
2. Obtain the milk carton filled with frozen water. Remove the carton from the block of ice. Calculate and record the overall volume of the block and the area of one side of the ice block.

3. Add the sand and small pebbles to one end of the pan so that they cover about half the area of the pan sloping toward the middle. Elevate this end of the pan using the wooden block as shown in Figure 15.2.

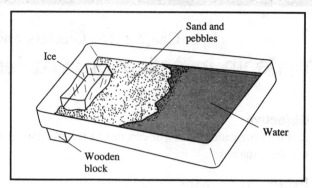

Figure 15.2

4. Slowly add water to the opposite end of the pan. Be sure that the water does not cover the sand, but only touches the edge of it. Measure and record the depth of the water at the deepest point. Then measure and record the distance from the end of the pan covered with sand to the point where the sand touches the water.

5. Place the block of ice lengthwise in the pan on top of the sand as shown in Figure 15.2. What percentage of the total area of the pan calculated in Step 1 is covered by ice?

6. As the ice begins to melt, pick up the ice block. Note the appearance of the bottom of the ice block and of the sand under the ice. What is happening to the ice block? What is happening to the sand under the ice block? Place the ice block back on the sand.

7. While the ice is melting, calculate the expected rise in the pan's water level using the following formula:

$$\text{rise in water level (cm)} = \frac{\text{volume of water in ice block (cm}^3)}{\text{area of pan covered with water (cm}^2)}$$

8. When the ice is completely melted, measure and record the depth of water at the deepest point in the pan. What is the difference in the water level after the ice has melted in the pan? How does this compare with the value you calculated in Step 7? Explain any difference.

9. Remeasure and record the distance from the end of the pan covered with sand to the point where the sand touches the water. What is the difference in this distance after the ice has melted in the pan?

Analysis and Conclusions

1. How is the ice block model different from a real glacier on the earth?
2. How does this model represent what would happen on the earth if the Antarctic ice sheet melted?
3. In this investigation, you used a physical model to simulate an occurrence in nature. How else do scientists use models? What kinds of errors may occur when using models?

Extension

The total area of the earth is approximately 511,000,000 km². About 70% of the earth's surface is covered with water. The volume of water locked up as ice in the Antarctic ice sheet is approximately 17,900,000 km³. Calculate the area of the earth, in square kilometers, that is covered with water. Then use the following equation to find the average worldwide rise in sea level that would occur if the Antarctic ice sheet melted. Convert your answer to meters.

$$\text{rise in sea level (km)} = \frac{\text{volume of water in Antarctic ice (km}^3)}{\text{area of earth covered with water (km}^2)}$$

How is the mathematical model for the rise in sea level similar to the physical model used in this investigation?

Chapter 15: Glaciers and Erosion
In-Depth Investigation: Glaciers and Sea Level

Objective
In this investigation, you will construct a model to simulate what would happen if the Antarctic ice sheet melted.

Prelab Preparation
What is the area of Figure 15.1a? What is the volume of Figure 15.1b?

_____

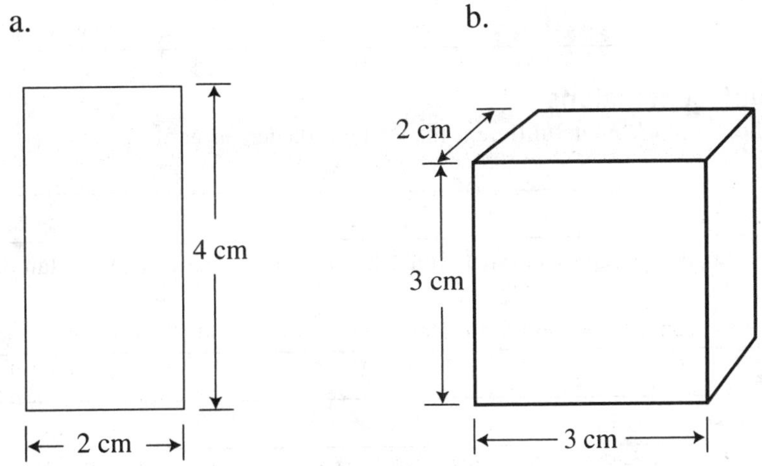

a.   b.

Figure 15.1

Observations
1. What is the surface area of the bottom of the pan?

_____

2. What is the volume of the ice block? What is the area of one side?

_____

3. What is the depth of water at the deepest point in the pan?

_____

4. What is the distance from the end of the pan covered with sand to the point where the sand touches the water?

_____

5. What percentage of the total area of the pan is covered by ice?

_____

6. a. As the ice begins to melt, what happens to the bottom of the ice block?

_____

b. What happens to the sand under the ice block?

_____

7. What is the calculated rise in water level in the pan?

_____

8. a. After the ice is completely melted, what is the depth of water in the pan?

_____

   b. What is the difference in the water level after the ice has melted?

_____

   c. How does this compare with the value you calculated in Step 7? Explain any differences.

_____

_____

9. a. After the ice has melted, what is the distance from the end of the pan covered with sand to the point where the sand touches the water?

_____

   b. What is the difference in this distance after the ice has melted?

_____

Analysis and Conclusions

1. How is the ice block model different than a real glacier on earth?

_____

_____

2. How does this model represent what would happen on the earth if the Antarctic ice sheet melted?

_____

_____

_____

3. During this investigation, you used a physical model to simulate an occurence in nature. How else do scientists use models? What kinds of errors may occur when using models?

_____

_____

_____

_____

Extension

What is the area of the earth, in square kilometers, that is covered with water?

_____

What is the average worldwide rise in sea level, in kilometers, if the Antarctic ice sheet melted? Convert this answer to meters.

_____

How is the mathematical model for the rise in sea level similar to the physical model used earlier in this investigation?

_____

_____

M O D E R N   E A R T H   S C I E N C E

Chapter 16: Erosion by Wind and Waves
In-Depth Investigation: Beaches

Objective
In this investigation, you will examine a model showing how the forces generated by wave action build up, shape, and wear away beaches.

Skills
observing, predicting, interpreting a model

Part I Introduction
According to some estimates, 50% of the population in the United States lives within 50 miles of a shoreline. Beaches provide a source of income for many residents. Coastal management has become a growing concern as beaches are increasingly used for resources and recreation. The supply of sand for most beaches has been cut off by dams built on rivers and streams that would otherwise carry sand to the sea. Waves generated by storms also continuously wash away beaches.

Materials
masking tape	two milk cartons (small)	water
sand (5–6 lb.)	wooden block (large)	pebbles
rocks (small)	plastic container	plaster of Paris
stream table	metric ruler	

Prelab Preparation
1. Review Chapter 16, Section 16.2 Wave Erosion, pages 302–306.
2. One day before you begin the investigation, make two plaster blocks. Mix a small amount of water with the plaster of Paris until the mixture is smooth. Before you pour the plaster mixture into the milk cartons, add five or six small rocks to the mixture for added weight. Let the plaster harden overnight. Carefully peel off the milk carton.

Procedure
1. Prepare a stream table or other similar large, shallow container. Make a beach by placing a mixture of sand and small pebbles at one end of the container. The beach should occupy about one fourth of the length of the container.
2. In front of the sand, add water to a depth of 2 to 3 centimeters. What happened to the beach when water was first poured into the container?
3. Using the large wooden block, generate several waves by moving the block up and down in the water at the end of the container opposite the beach. See Figure 16.1. Continue this wave action until about half the beach has shown some movement. Describe the beach after this wave action has taken place. What happened to the particles of fine sand?
4. Predict what will happen to the beach if it has no source of additional sand.

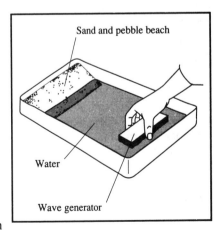

Sand and pebble beach

Water

Wave generator

Figure 16.1

5. Remove the sand and rebuild the beach.

6. In some places, breakwaters have been built off-shore to protect beaches from washing away. Build a breakwater by placing two plaster blocks across the middle of the container. Leave a 4-cm space between the blocks. See Figure 16.2.

7. Use a wooden block to generate waves as in Step 1. Record your observations.

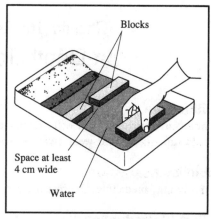

Figure 16.2

Part II Introduction

Most waves that reach a beach move straight toward the beach. However, some waves in deeper water move at an angle to the beach. As these waves approach the beach at an angle, they set up a current called a long-shore current. A longshore current moves parallel to the beach and in the same direction that the waves were moving in deep water.

Procedure

8. Drain the water and make a new beach along one side of the container for about half its length. See Figure 16.3 Predict what effect a longshore current would have on sand just offshore. How would this affect sand on the beach?

9. Using the wooden block, generate a series of waves from the same end of the container as the end of the beach. See Figure 16.3. Record your observations. What happened to the beach? What happened to the shape of the waves along the beach?

10. Rebuild the beach along the same side of the container.

11. A jetty or dike can be built out into the ocean to intercept and break up a longshore current. Place one of the small plaster blocks in the sand to act as a jetty. See Figure 16.4.

12. As you did before, use the wooden block to generate waves. Describe the results.

13. Remove the wet sand and place it in a container. Dispose of the water. *NOTE: Follow your teacher's instructions for disposal of the sand and water. Never pour water containing sand into a sink.*

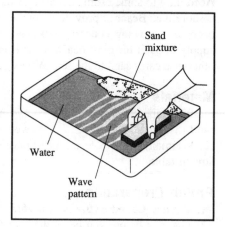

Figure 16.3

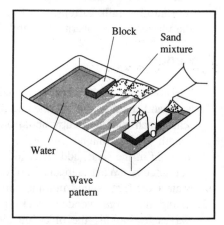

Figure 16.4

Analysis and Conclusions

1. How does wave action build up a beach? How does wave action wear away a beach?

2. How do longshore currents change the shape of a beach?

3. What effect would a series of jetties have on a beach?

Extensions

1. What can be done to preserve a recreational beach area from erosion as a result of excessive use by people?

2. What can be done to preserve a recreational beach area from being washed away as a result of wave action and longshore currents?

M O D E R N   E A R T H   S C I E N C E

Chapter 16: Erosion by Wind and Waves
In-Depth Investigation: Beaches

Objective
In this investigation, you will examine a model showing how the forces generated by wave action build up, shape, and wear away beaches.

Part I Observations
1. What happened to the beach when water was first poured into the container?

_____

_____

2. a. Describe the appearance of the beach after the wave action has taken place.

_____

_____

   b. What happened to the particles of fine sand?

_____

_____

3. What will happen to the beach if it has no source of additional sand?

_____

_____

4. What happened to the beach when you generated waves in front of a breakwater?

_____

_____

_____

_____

Part II Observations
5. a. What effect would a longshore current have on sand just offshore?

_____

   b. How would this affect the sand on the beach?

_____

_____

6. What happened when you generated waves after placing a jetty in the sand?

_____

_____

Analysis and Conclusions
1. a. How does wave action build up a beach?

_____

_____

b. How does wave action wear away a beach?

_____

_____

2. How do longshore currents change the shape of a beach?

_____

_____

3. What effect would a series of jetties have on a beach?

_____

_____

_____

Extensions

1. What can be done to preserve a recreational beach area from erosion as a result of excessive use by people?

_____

_____

_____

2. What can be done to preserve a recreational beach area from being washed away as a result of wave action and longshore currents?

_____

_____

_____

_____

M O D E R N   E A R T H   S C I E N C E

Chapter 17: The Rock Record
In-Depth Investigation: Fossils

Objective

In this investigation, you will use various methods to make models of trace fossils.

Skills

making and interpreting models, inferring

Introduction

Paleontologists study fossils to find evidence of the kinds of life and conditions that existed on the earth in the geologic past. Fossils are the remains of ancient plants and animals or evidence of their presence. Some fossils are the preserved or altered bodies of organisms. Others, called trace fossils, such as footprints, tracks, burrows, and borings provide indirect evidence of what an ancient animal looked like. A mold is formed when an animal or plant is buried in sediments that later harden into rock. In time, the body of the organism decays, leaving an empty space, or mold, in the rock with the same shape and surface markings as the organism. Under certain conditions, the mold may fill with minerals that harden to produce a replica of the outer surface of the original organism. This mineral replica is called a cast. Another type of fossil is an imprint, such as the thin impression made by a fish or a leaf. Carbon films on rock surfaces left by the decayed soft parts of an organism, such as a leaf, are examples of impressions. The shape and some surface features of the organism are visible in the carbon film.

Materials

hard objects (shell, key, paper clip, etc.)	plastic container	tweezers
	plastic spoon	sheet of white paper
modeling clay	pencil or wooden dowel 15 cm)	water
newspaper	soft carbon paper	lab apron
plaster of Paris	leaf	

Prelab Preparation

1. Review Chapter 17, Section 17.3 The Fossil Record, pages 287–289.
2. Describe three ways in which fossils can be classified based on how they were formed.
3. Why are coal, oil, and gas called fossil fuels?

Procedure

1. Place a ball of modeling clay on a flat surface. Press the clay down to form a flat disk about 8 cm in diameter. Turn the clay over so that the smooth, flat surface is facing up.
2. Choose a small hard object. Press the object onto the clay surface so that it leaves an indentation in the clay. Remove the object from the clay carefully so that you do not disturb the indentation. Is the indentation left by the object a mold or a cast? What features of the object are best shown in the indentation? Sketch the indentation.
3. Fill a plastic container with water to a depth of 1 to 2 cm. Stir in enough plaster of Paris to make a paste with the consistency of whipped cream.
4. Spread several sheets of newspaper on a flat surface. Place the clay on the newspaper.
5. Using the plastic spoon, fill the indentation with plaster. Allow excess plaster to run over the edges of the imprint. Let the plaster set for about 15 minutes until it hardens.

6. On a second piece of smooth, flat clay, make a shallow imprint to represent the burrow or footprint of an animal. Fill the model trace fossil with plaster and let it harden.

7. After the plaster has hardened, remove both pieces of plaster from the clay. Do the pieces of hardened plaster represent molds or casts? Sketch your fossil imprint.

8. Place the carbon paper carbon side up on a flat surface. Gently place the leaf on the carbon paper and cover it with several sheets of newspaper. Roll the pencil or wooden dowel back and forth across the surface of the newspaper several times, pressing firmly to bring the leaf into solid contact with the carbon paper.

9. Remove the newspaper. Lift the leaf using the tweezers and place it carbon side down on a clean sheet of white paper. Cover the leaf with clean newspaper and roll your pencil across the surface of the paper once again, as in Step 8.

10. Remove the newspaper and leaf. Observe and describe the carbon-print left by the leaf.

Analysis and Conclusions

1. Look at the molds and casts made by others in your class. Identify as many of the objects used as you can.

2. How does the carbon-print you made differ from an actual carbon-print trace fossil? Why is a carbon-print a trace fossil?

3. Why are carbon-prints, molds, and casts trace fossils?

Extension

Look at the organisms shown in Figure 17.1. Which of these organisms would be most likely to form fossils? Which would leave trace fossils? Explain.

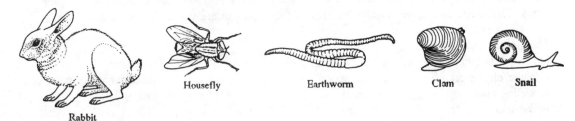

Rabbit   Housefly   Earthworm   Clam   Snail

Figure 17.1

M O D E R N   E A R T H   S C I E N C E

Chapter 17: The Rock Record
In-Depth Investigation: Fossils

Objective
In this investigation, you will use various methods to make models of trace fossils.

Prelab Preparation
1. Describe three ways in which fossils can be classified based on how they were formed.

_____

_____

2. Why are coal, oil, and gas called fossil fuels?

_____

_____

Observations
1. **a.** Is the indentation left by the object a mold or a cast?

_____

b. What features of the object are best shown in the indentation?

_____

c. Sketch the indentation in the space below.

┌───┐
│   │
│   │
│   │
│   │
│   │
│   │
└───┘

2. **a.** Do the pieces of hardened plaster represent molds or casts?

_____

b. Sketch your fossil imprint in the space below.

┌───┐
│   │
│   │
│   │
│   │
│   │
└───┘

Analysis and Conclusions

1. List all objects that you can identify from the molds and casts.

_____

_____

_____

_____

2. How does the carbon-print you made differ from an actual carbon-print trace fossil?

_____

_____

_____

3. Why are carbon-prints, molds, and casts trace fossils?

_____

_____

_____

Extensions

1. Which of the organisms in Figure 17.1 would be most likely to form fossils? Explain.

_____

_____

_____

2. Which of the organisms in Figure 17.1 would leave trace fossils? Explain.

_____

_____

_____

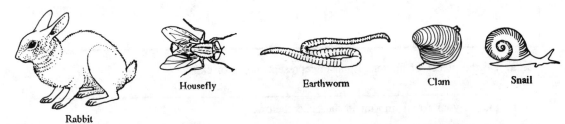

Rabbit   Housefly   Earthworm   Clam   Snail

Figure 17.1

M O D E R N   E A R T H   S C I E N C E

Chapter 19: The History of the Continents
In-Depth Investigation: History in the Rocks

Objective
In this investigation, you will discover how the geologic history of an area can be determined by examining the arrangement of fossils and rock layers.

Skills
analyzing, classifying, comparing, contrasting, interpreting data, observing

Introduction
Geologists have discovered much about the geologic history of North America by studying the arrangement of fossils in rock layers, as well as the arrangement of the rock layers themselves. Fossils provide clues about the environment during the time of their existence. Scientists can tell the age of the rocks in which the fossils are found because the ages of many of these fossils have been determined by radioactive dating. The information obtained by radioactive dating, fossil age, and rock arrangement helps to determine if any changes have occurred in the arrangement of the rock layers through geologic time.

Materials
paper
pencil

Prelab Preparation
1. Review Chapter 17, Section 17.1 Determining Relative Age, pages 323–326, and Section 17.3 The Fossil Record, pages 334–339. Review Chapter 19, Section 19.3 Formation of the Grand Canyon, pages 374–377.
2. Define the terms *index fossil* and *law of superposition* as they relate to the study of rock layers.

Procedure
1. Study the index fossils shown in Figure 19.1 on page 90. Note their placement in related groups along with the geologic periods in which they lived.
2. Select one of the four fossil arrangements in Figure 19.2 on page 90. Figure 19.2 shows how some of these fossils might be found in a series of rock layers. Record the number of the arrangement you are using.
3. Using Figure 19.1, identify all the fossils in your arrangement and the geologic time in which they lived.
4. List the fossil names in order from oldest to youngest.
5. Do the fossils in your arrangement appear in the correct order of their geologic times?
6. Do the fossils in your arrangement show a complete sequence of geologic periods with none missing? If not, which periods are missing?
7. Select another arrangement showing a different group of fossils. Repeat Steps 2–6 with this new group. Complete all four arrangements.

Analysis and Conclusions
1. What processes or events might explain the order in which each of the fossil arrangements were found?

2. Based on your observations in the procedure, why is it necessary that a fossil be found in a wide variety of geographic areas in order to be considered an index fossil?

3. Study Arrangement 3 in Figure 19.2. Note that there is a rock layer containing no fossils in between two rock layers that contain fossils. How might this have occurred?

Extensions

1. Collect fossils found in your area. Identify the fossils you have collected and describe what your area was like when the organisms existed.

2. Find out what types of sedimentary rock usually have the most fossils.

3. How are index fossils used to help petrologists locate oil deposits?

Fossils in Geologic Time

Geologic Periods	Name of Animal Group				
	Brachiopoda	Echinodermata	Mollusca	Arthropoda	Chordata
Quaternary			Pelecypod / Pelecypod		Mammal
Tertiary			Gastropod		Mammal / Shark Tooth
Cretaceous	Brachiopod	Echinoid	Cephalopod		
Triassic			Cephalopod		
Pennsylvanian	Brachiopod				
Mississippian	Brachiopod	Crinoid / Blastoid	Cephalopod		
Devonian	Brachiopod			Trilobite	
Silurian	Brachiopod				
Ordovician			Cephalopod	Trilobite	

Figure 19.1

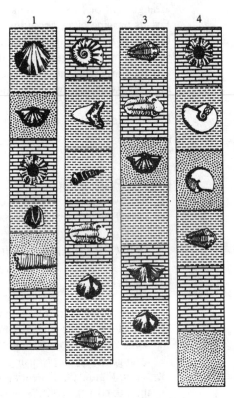

Figure 19.2

M O D E R N   E A R T H   S C I E N C E

Chapter 19: The History of the Continents
In-Depth Investigation: History in the Rocks

Objective
In this investigation, you will discover how the geologic history of an area can be determined from the arrangement of fossils and rock layers.

Prelab Preparation
Define the terms *index fossil* and *law of superposition* as they relate to the study of rock layers.

_____

_____

_____

_____

Observations
1. List the fossil names from oldest to youngest in each arrangement.

_____

_____

_____

_____

_____

_____

2. In the four arrangements shown in Figure 19.2, do the fossils appear in the correct order of their geologic times? Name fossils, if any, that appear out of order in each arrangement.

_____

_____

_____

_____

3. Do the fossils in the four arrangements represent complete sequences of geologic periods? If not, which periods are missing in each arrangement?

_____

_____

_____

_____

_____

_____

_____

Analysis and Conclusions

1. What processes or events might explain the order in which each of the fossil arrangements were found?

   _____

   _____

2. Based on your observations in the procedure, why is it necessary that a fossil be found over a wide geographic area in order to be considered an index fossil?

   _____

   _____

   _____

   _____

3. Give a few possible reasons why there is a rock layer in Arrangement 3 containing no fossils in between rock layers that contain fossils.

   _____

   _____

   _____

   _____

Extensions

1. Collect fossils found in your area. Identify the fossils you have collected and describe what your area was like when the organisms existed.

   _____

   _____

   _____

2. What types of sedimentary rock usually have the most fossils?

   _____

   _____

   _____

   _____

   _____

3. How do index fossils aid in the discovery of oil?

   _____

   _____

   _____

   _____

M O D E R N   E A R T H   S C I E N C E

Chapter 20: The Ocean Basins
In-Depth Investigation: Ocean-Floor Sediments

Objective
In this investigation, you will determine the relationship between the size of sediment particles and their settling rate in water.

Skills
comparing and contrasting, interpreting data, observing

Introduction
Most of the ocean floor is covered with a layer of sediment that varies from 0.3 to 0.5 km in thickness. Much of this sediment is thought to have originated on land through the process of weathering. Through erosion, the sediment has made its way to the deep ocean basins.

Several factors determine where the sediment carried to the ocean will be deposited. One factor is the size of the particles that reach the ocean. By using soil samples of given particle size, you will determine the relationship between the size of particles and their settling rate in water. As you carry out this investigation, watch for any other factors which may affect settling rate.

Materials
1 lb. dry mixed sand/soil	paper cups	stopwatch
clear plastic column (32" × 1.5")	sieves (4 mm, 2 mm, 0.5 mm)	water
grease pencil	paper towels	measuring cup
teaspoon	rubber stopper	tape
ring stand and clamp		

Prelab Preparation
1. Review Chapter 20, Section 20.3 Ocean Basin-Sediments, pages 399–401.
2. On a surface protected by paper towels, prepare the mixed sand/soil samples. Stack the sieves on top of each other with the coarsest sieve on top and the finest on the bottom. Pour the mixed sand/soil sample into the top sieve, and sift it through to separate particles in the following size ranges: *Coarse*—the particles that remain on the top of the 4-mm sieve; *Medium*—the particles that pass through the 4-mm sieve but remain on the 2-mm sieve; *Medium fine*—the particles that pass through the 2-mm sieve but remain on the 0.5-mm sieve; *Fine*—the particles that pass through the 0.5-mm sieve. Place each sample in a paper cup.

Procedure
1. Plug one end of the plastic column with a rubber stopper, and secure the stopper to the column with tape. Carefully fill the column with water to a level about 5 cm from the top, and place the column in a vertical position using the ring stand and clamp. Allow the water to stand until all large air bubbles have escaped. With the grease pencil, mark the water level on the column. This will be the starting line. Next, draw a line about 5 cm from the bottom of the column. This will be the finish line.

2. Have a member of your lab group dump one teaspoon of the coarse sample into the water column. The other group member should record two time measurements as follows:
   a. Using a stopwatch, start timing when the first particles hit the start line on the column, and stop timing when they reach the finish line. Repeat this procedure twice. Record the time for each trial in Table 20.1.
   b. Next, use the stopwatch to determine how long it takes the last particle in the sample to travel from the start line to the finish line at the base of the column. Repeat this procedure twice. Record the time for each trial in Table 20.1.
3. Determine the average time of the three trials for the first measurement. Do the same for the second measurement. Record the averages in Table 20.1.

4. How long did it take for the first particles in the coarse sample to reach the finish line? How long did it take for all the coarse particles to reach the finish line?
5. Pour the sand/soil and water from the column into the container provided by your teacher. *Note: Do not pour sand or soil into the sink.*

Table 20.1

Soil samples	Trial 1	Trial 2	Trial 3	Average
Coarse	First time measurement:			
	Second time measurement:			
Medium	First time measurement:			
	Second time measurement:			
Medium-fine	First time measurement:			
	Second time measurement:			
Fine	First time measurement:			
	Second time measurement:			

6. Refill the plastic column with water up to the original level marked with the grease pencil.
7. Repeat Steps 1, 2, 3, and 5 for the remaining sample sizes. What were the settling times for the medium particles? the medium-fine particles? the fine particles? Record these measurements and the averages in Table 20.1.
8. Refill the plastic column with water. Pour 20 g of unsieved soil into the column and allow it to settle for five minutes. After five minutes, look at the column. Do layers of sediment appear in the column? Why does the water remain slightly cloudy even after most of the particles have settled?

Analysis and Conclusions
1. Compare the settling time of the medium particles with the settling time of the medium-fine particles.
2. Do similar-sized particles fall at the same rate?
3. Other than size, what factors would you expect to influence how rapidly particles fall in sea water?
4. How do the results in Step 8 help to explain why the deep ocean basins are covered with a very fine layer of sediment while areas near the shore are covered with coarse sediment?

Extension
What are some sources of the sediment that is found in the ocean?

M O D E R N   E A R T H   S C I E N C E

Chapter 20: The Ocean Basins
In-Depth Investigation: Ocean-Floor Sediments

Objective
In this investigation, you will determine the relationship between the size of sediment particles and their settling rate in water.

Observations
Record the time measurements in Table 20.1. Calculate and record the averages.

Table 20.1

Soil samples	Trial 1	Trial 2	Trial 3	Average
Coarse	First time measurement:			
	Second time measurement:			
Medium	First time measurement:			
	Second time measurement:			
Medium-fine	First time measurement:			
	Second time measurement:			
Fine	First time measurement:			
	Second time measurement:			

1. What is the average settling time recorded for the first particles in the coarse sample to reach the finish line? What is the average settling time for all the coarse particles to reach the finish line?

   _____

   _____

2. What are the average settling times recorded for the medium-sized particles? The medium-fine–sized particles? The fine particles?

   _____

   _____

3. Do layers of sediment appear in the cylinder in Step 8?

   _____

4. Why does the water remain slightly cloudy even after most of the particles have settled in Step 8?

   _____

   _____

   _____

Analysis and Conclusions

1. How does the settling time of the medium particles compare with the settling time of the medium-fine particles?

   _____

   _____

2. Do similar-sized particles fall at the same rate?

   _____

3. Other than size, what factors would you expect to influence how rapidly particles fall in sea water?

   _____

   _____

4. How do the results in Step 8 of the procedure help to explain why the deep ocean basins are covered with a very fine layer of sediment while areas near the shore are covered with coarse sediment?

   _____

   _____

   _____

Extension

What are some sources of the sediment that is found in the ocean?

_____

_____

_____

M O D E R N   E A R T H   S C I E N C E

Chapter 21: Ocean Water

In-Depth Investigation: Ocean Water Density

Objective
In this investigation, you will observe the effects of temperature and salinity on the density of salt water.

Skills
measuring, observing, comparing, determining cause and effect, inferring

Introduction
The density of ocean water varies in different areas of the ocean. Density differences are affected by the amount of dissolved solids, including salt, and the temperature of the water. Because salt is the most abundant dissolved solid in the ocean, a change in salinity will change the water's density. The salinity of an area of the ocean is also affected by the rate of evaporation or freezing and by the amount of fresh water and salts added by rivers and glacial runoff. The temperature of the ocean is determined by the amount of infrared radiation it receives.

Materials
plastic straw	modeling clay	beaker (250-mL)
grease pencils (yellow and red)	table salt	Bunsen burner
50-mL graduated cylinder	ring stand and clamp	teaspoon
heat-resistant gloves	wire gauze	safety goggles
thermometer (Celsius)	metric ruler	scissors
freezer	water	

Prelab Preparation
1. Read Chapter 21, Section 21.1 Properties of Ocean Water, pages 407–412.
2. Review the safety guidelines for fire, heating, and eye and hand safety.
3. Define density.
4. Constructing a hydrometer:
    a. Hold your index finger over one end of a plastic straw and slowly press the open end into a piece of clay until the straw is filled with about 5 cm of clay.
    b. Place the straw, clay end down, in a graduated cylinder filled with 50 mL of tap water. (If the straw does not float upright, cut off the open end at 1-cm intervals until it does.) Use a red grease pencil to mark the water level on the straw. Remove the straw from the water and draw a continuous line around the straw at the mark.
    c. Using a yellow grease pencil, draw lines around the straw at 1-cm intervals above and below the red line. The red line will be used as a reference point.

Procedure
1. Pour 100 mL of room-temperature tap water into a beaker. Stir in 2 teaspoons of table salt until it is all dissolved. Measure and record the water temperature in Table 21.1.
2. Place the hydrometer in the salt water. In Table 21.1, record the density by counting the marks above the red line to the water's surface.
3. Set up the ring-stand apparatus and Bunsen burner as shown in Figure 21.1 on page 98. **CAUTION: Wear you safety goggles.**
4. Light the Bunsen burner. Keep the flame low. Place the beaker of salt water on the wire gauze. **CAUTION: Wear heat-resistant gloves.** Hold a thermometer in the water. Do not let the thermometer touch the bottom of the beaker.

5. When the water's temperature reaches 25°C, turn off the Bunsen burner. Immediately place the hydrometer in the water. Record the relative density in Table 21.1.

6. Repeat Steps 4 and 5, heating the water until it reaches 30°C. As the temperature increases, does the water's density increase or decrease?

7. Heat the salt water until it begins to boil. Continue to boil the water for 5 minutes. Turn off the burner. Place the hydrometer in the water. Measure and record the water's density in Table 21.2.

8. Boil the water for another five minutes. Measure and record the water's density.

9. Repeat Step 8. Did the density of the water increase or decrease with each boiling?

10. Originally there was 100 mL of water in the beaker. How much water is in the beaker now?

Table 21.1

Temperature (°C)	Density (cm above or below the red line)
25	
30	

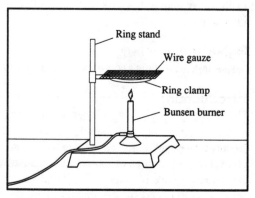

Figure 21.1

Analysis and Conclusions

1. Explain how and why warming the water affected its density.

2. Based on your observations, infer what the density of polar ocean waters would be compared with the density of equally saline water near the equator. Explain your answer.

3. Why did the amount of water in the beaker change? Explain why boiling the water affected its density.

Table 21.2

Minutes of boiling	Density (cm above or below the red line)
5	
10	
15	

Extension

Place a beaker of salt water in a freezer until a crust of ice forms at the top. Break up and remove the ice from the water. Is the water denser or less dense than before freezing? Explain why.

M O D E R N   E A R T H   S C I E N C E

Chapter 21: Ocean Water
In-Depth Investigation: Ocean Water Density

Objective
In this investigation, you will observe the effects of temperature and salinity on the density of salt water.

Prelab Preparation
Define density.

_____

_____

Observations
1. Record the temperature of the salt water in Table 21.1.
2. Record the density of the water at room temperature, 25°C, and 30°C in Table 21.1.
3. Record the density of the water after 5, 10, and 15 minutes of boiling in Table 21.2.

Table 21.1

Temperature (°C)	Density (cm above or below the red line)
25	
30	

Table 21.2

Minutes of boiling	Density (cm above or below the red line)
5	
10	
15	

4. As the temperature increases, does the water's density increase or decrease?

_____

5. Did the density of the water increase or decrease with each boiling?

_____

6. Originally there was 100 mL of water in the beaker. How much is in the beaker now?

_____

Analysis and Conclusions

1. Explain how and why warming the water affected its density.

   _____

   _____

   _____

2. Based on your observations, infer what the density of polar ocean water would be compared with the density of water of equal salinity near the equator. Explain your answer.

   _____

   _____

   _____

   _____

3. Why did the amount of water in the beaker change? Explain why boiling the water affected its density.

   _____

   _____

   _____

   _____

   _____

   _____

Extension

Place a beaker of salt water in a freezer until a crust of ice forms at the top. Break up and remove the ice from the water. Is the water denser or less dense than before freezing? Explain.

   _____

   _____

   _____

M O D E R N   E A R T H   S C I E N C E

Chapter 22: Movements of the Ocean
In-Depth Investigation: Wave Motion

Objective
In this investigation, you will simulate wave motion to observe how energy generates wave motion in water. You also will observe the properties of waves.

Skills
observing, measuring, estimating, analyzing, experimenting

Introduction
The source of wave movement in water is energy, which is generated primarily from wind. To a person standing on a beach and watching the waves come in to the shore, it may not be easy to understand that wind is the primary energy source of waves. Waves appear to move forward because of the actual movement of the water. However, only the energy of the waves moves forward; the water moves very little. In this investigation, you will observe how energy generates wave movement.

Materials
graph paper	sheet of paper (2 × 1 m)	3 colored pens or pencils
marker	thin rope (2.5 m)	2 cloth ties (about 50 cm in length)
meter stick		

Prelab Preparation
1. Review Chapter 22, Section 22.2 Ocean Waves, pages 433–438.
2. Study Graph 22.1. Label the wave crests and wave troughs. The wavelength is the distance between two successive crests or troughs. What is the wavelength? What is the wave height?

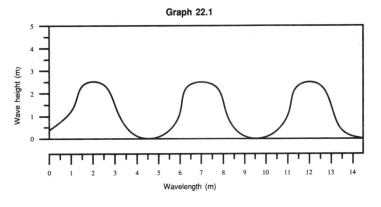

Graph 22.1

3. The wave period is the time required for two successive crests or troughs to pass a certain point. Wave speed can be calculated using the following formula:

$$\text{Wave speed} = \frac{\text{wavelength}}{\text{wave period}}$$

What is the speed of waves in Graph 22.1 if the wave period is five seconds?

Procedure
1. Work with two partners for this investigation. Begin by tying one end of the rope to the leg of a chair or table. Be sure it is tied securely.
2. On the large sheet of paper, use the meter stick to draw a grid similar to the one shown in Figure 22.1 on page 102. Draw and label the grid using the measurements in Figure 22.1.

3. Place the sheet of paper on the floor and line up the rope along the 2-m line of the grid.

4. To make waves, you or one of your partners must move the free end of the rope from side to side as shown in Figure 22.2. *Note: Be sure to maintain a constant motion with the rope.*

5. While one person is moving the rope, another person must mark on the paper where a crest of a wave hits. The third group member must mark on the paper where a trough of a wave hits.

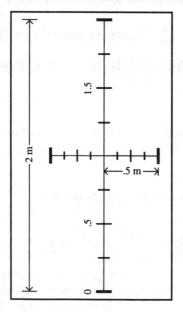

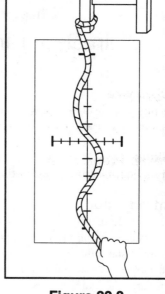

Figure 22.1   **Figure 22.2**

6. On Graph 22.2, page 104, plot a wave that represents the wave that you observed in Step 5. Begin at 0 in the middle of the graph, and plot the height and the length. Indicate the direction of the wave's motion.

7. Continue the investigation by changing the motion of the rope. Move the rope at a fast speed. *Note: Be sure to move the free end of the rope the same distance from side to side.*

8. As soon as a constant motion has been established, repeat Steps 5 and 6. Using a different colored pen or pencil, plot one of these waves on Graph 22.2.

9. Next, generate very small waves. Repeat Steps 5 and 6 using a third color.

10. On Graph 22.2, label a crest and a trough on each of the waves that you have plotted. What are the wavelengths of the three waves you plotted? What are their wave heights? Using the formula in the prelab preparation, calculate the wave speeds of the three waves represented on the graph if each wave period is six seconds.

11. Tie the two pieces of cloth around the middle of the rope about 15 cm apart. Set up another wave motion by moving the end of the rope. Observe and record the motion of the cloth ties compared with the motion of the waves.
the motion of the waves.

Analysis and Conclusions

1. How do the wave motions differ on your graph? If these were real water waves, what might be the cause(s) of the different motions?

2. How is the action of the rope similar to wave movement in water?

3. How do the motions of the cloth ties differ from the wave's motion?

4. What do the motions of the cloth ties tell you about wave movement in water?

Extension

Repeat the investigation using a 4-m rope. Construct a graph similar to Graph 22.2 but extend the x-axis to plot a 4-m length. Observe and plot five waves of varying speeds and heights. Compare waves generated on a 2-m rope to those generated on a 4-m rope. Describe your results. Calculate wave speeds, using six seconds as the wave period for each wave that you plotted.

M O D E R N   E A R T H   S C I E N C E

Chapter 22: Movements of the Ocean
In-Depth Investigation: Wave Motion

Objective
In this investigation, you will simulate wave motion in order to observe how energy generates wave motion on water. You will also observe various properties of waves.

Prelab Preparation
1. Label the wave crests and wave troughs on Graph 22.1.

Graph 22.1

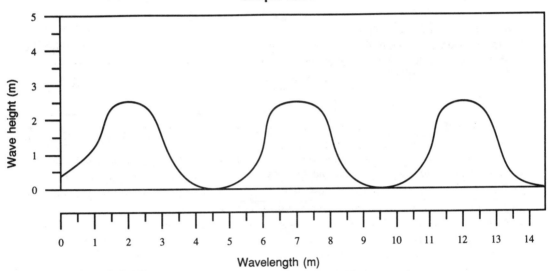

2. **a.** What is the wavelength in Graph 22.1?

   _____

   b. What is the wave height?

   _____

   c. What is the speed of the waves in Graph 22.1 if the wave period is five seconds?

   _____

Observations
1. Plot the waves you simulated with the rope on Graph 22.2, page 104.

2. **a.** What are the wavelengths of the three waves you plotted?

   _____

   b. What are the wave heights of the three waves?

   _____

   c. What are the wave speeds of the three waves if each wave period is six seconds?

   _____

Graph 22.2

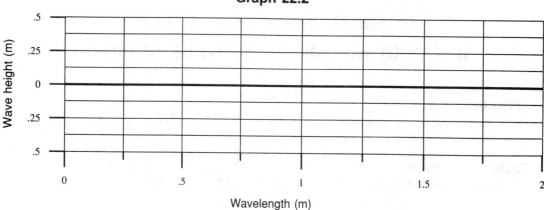

Analysis and Conclusions

1. How do the wave motions differ on your graph? If these were real water waves, what might be the cause(s) of the different motions?

_____

_____

_____

_____

2. How is the action of the rope similar to wave movement in water?

_____

_____

_____

_____

3. How do the motions of the cloth ties differ from the wave's motion?

_____

_____

4. What do the motions of the ties tell you about wave motion in water?

_____

_____

_____

Extension

Describe the five waves you plotted on your graph. Compare the waves plotted using a 4-m rope to the waves plotted using a 2-m rope. Are there differences and/or similarities? Explain. What are the wave speeds that you calculated for each wave?

_____

_____

_____

M O D E R N   E A R T H   S C I E N C E

Chapter 23: The Atmosphere

In-Depth Investigation: Air Density and Temperature

Objective
In this investigation, you will observe how the density of air is affected by changes in its temperature.

Skills
observing, recording data, inferring, predicting

Introduction
Circulation in the earth's atmosphere is caused by the sinking of cold, dense air and the rising of warm, less dense air. The uneven heating of air causes this rising and sinking. The vertical movement of air caused by uneven heating is called convection. Convection is the main cause of air circulation in the earth's atmosphere.

Materials
400 mL beaker
disposable syringe (60 cc)
cold water

hot water
glycerine
petroleum jelly

ice
Celsius thermometer
graph paper

Prelab Preparation
1. Review Chapter 23, Section 23.1 Characteristics of the Atmosphere, pages 455–462, and Section 23.3 Winds, pages 469–473.
2. Density (D) is defined as the mass (m) per unit volume (V) of a substance, or $D = m/V$. A block of lead has a mass of 910 g and a volume of 80 cm^3. What is lead's density?
3. If the mass of a given substance remains the same while its volume increases, would the density of the substance increase or decrease?

Procedure
1. Remove the cap from the end of the syringe. Move the plunger until the cylinder is about two-thirds full of air. Add a dab of petroleum jelly to the end of the syringe and replace the cap. Gently pull on the plunger and release. Read the volume of air in the syringe. Record your answer in Table 23.1.

Table 23.1

Temperature (°C)	Volume (cm³)		
	Pull	Push	Average

2. With your finger on the cap, gently push on the plunger and release. Read the volume of air in the syringe and record your answer in Table 23.1. **CAUTION: Wear your safety goggles.**

3. Take the average of the two volume readings. This number should be close to the volume at room pressure and temperature. Measure the room temperature, and record both the average volume and temperature in Table 23.1.

4. Add ice to a beaker of cold water until the temperature is 5 to 10°C. Do not let the temperature drop below 5°C.

5. Place the syringe in the water so that the air in the cylinder is completely covered with water. After five minutes, the temperature of the air in the syringe will be the same as the water temperature.

6. Repeat Steps 1–3 to find the volume of air at the new temperature. Record your observations in Table 23.1.

7. Repeat Steps 5 and 6 with water temperatures of about 15°C, 40°C, and 60°C. Use hot tap water for the higher temperatures. If the plunger sticks at higher temperatures, lubricate the cylinder wall with a few drops of glycerine.

8. Plot your data in Graph 23.1. Draw a straight line through your data points. Some points may be slightly above or below this line. List reasons why all of the points do not fall on a straight line.

Analysis and Conclusions

1. According to your graph, what is the increase in volume, in cubic centimeters, as the temperature increases from 20°C to 40°C? As the temperature increases, does the density of the air in the syringe increase or decrease? Explain.

2. Based on your results, is hot air less dense or more dense than cold air? If air is warmed, does it become less dense or more dense? Would warm air tend to rise or sink in a body of colder air?

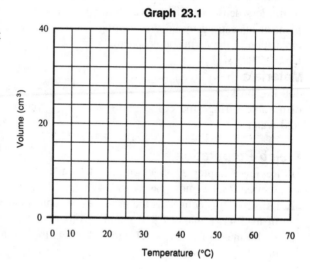

Graph 23.1

Extensions

1. When a temperature inversion occurs, a layer of warm air is trapped beneath a layer of cool air. How might this condition effect air quality?

2. A temperature inversion may last several days. What might cause conditions to return to normal?

M O D E R N   E A R T H   S C I E N C E

Chapter 23: The Atmosphere

In-Depth Investigation: Air Density and Temperature

Objective
In this investigation, you will observe how the density of air is affected by changes in its temperature.

Prelab Preparation
1. A block of lead has a mass of 910 g and a volume of 80 cm³. What is lead's density?
_____

2. If the mass of a given substance remains the same while its volume increases, will the density of the substance increase or decrease?
_____

Observations
1. Record your observations in Table 23.1.

Table 23.1

Temperature (°C)	Volume (cm³)		
	Pull	Push	Average

2. Plot your data in Graph 23.1 on page 108.
3. List reasons why the points in Graph 23.1 do not all fall on a straight line.
_____

Analysis and Conclusions
1. According to your graph, what is the increase in volume, in cubic centimeters, as the temperature increases from 20°C to 40°C.
_____

2. As the temperature increases, does the density of the air in the syringe increase or decrease? Explain.
_____
_____

Graph 23.1

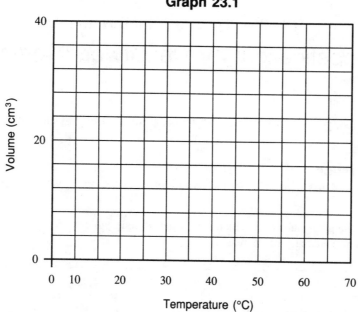

3. Is hot air less dense or more dense than cold air?

_____

4. If air is warmed, does it become less dense or more dense?

_____

5. Would warm air tend to rise or sink in a body of colder air?

_____

Extensions

1. How might a temperature inversion affect air quality?

_____

_____

2. What might cause conditions to return to normal?

_____

_____

M O D E R N   E A R T H   S C I E N C E

Chapter 24: Water in the Atmosphere

In-Depth Investigation: Relative Humidity

Objective

In this investigation, you will use wet-bulb and dry-bulb thermometer readings to determine relative humidity.

Skills

observing, measuring, reading a thermometer, inferring

Introduction

The earth's atmosphere acts as a storehouse for water that evaporates from the earth's surface. However, there is a limit to the amount of water vapor that the atmosphere can hold at a given temperature. When that limit is reached, the air is said to be saturated. Beyond this limit, no more water can be absorbed into the air. Air usually is not saturated with water vapor. Experiments show that air at a given temperature can never hold more than a certain amount of water vapor. For example, at normal room temperature (22°C) the air can hold about 20 grams of water per cubic meter. Relative humidity is a measure of how much water vapor is actually in the air compared with the maximum amount the air can hold at saturation. This comparison is expressed as a percentage. When the air holds all the water it can, it is said to have a relative humidity of 100%. Assuming that 1 m³ of air at 30°C could hold 30 g when saturated, you can express the relative humidity for the same air holding 20 g of water as

$$20 \text{ g}/30 \text{ g} = 0.67 = 67\%$$

If the air is not saturated, water can evaporate and enter the atmosphere. The energy needed to evaporate the water comes from the water itself. As the water evaporates, it loses heat energy. This loss of heat lowers the temperature of the water. A chart showing the cooling effect of evaporation can be used to determine relative humidity.

Materials

cotton cloth	ring stand with ring	Celsius thermometers (2)
plastic container	string	water

Prelab Preparation

1. Review Chapter 24, Section 24.1 Atmospheric Moisture, pages 480–483.
2. Study Graph 24.1 on page 110.
   a. How many grams of water can a cubic meter of air hold at a temperature of 20°C?
   b. How many grams of water can the air hold at 40°C?
3. If a cubic meter of air at 40°C holds 40 g of water, what is the relative humidity?

Procedure

1. Set up two thermometers as shown in Figure 24.1 on page 110. Wrap a piece of cotton cloth around the bulb of one thermometer. Adjust the length of the string so that only the cloth and not the thermometer bulb is immersed in the water. With this setup you can measure both the air temperature and the cooling effect of evaporation.
2. Make a prediction. Will the two thermometers have the same reading? If not, which thermometer will have the lower reading?

Graph 24.1

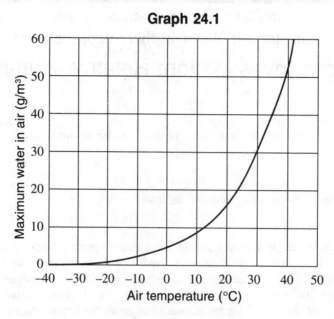

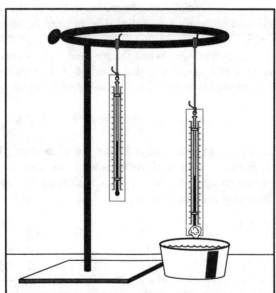

Figure 24.1

3. Fan both thermometers rapidly with a piece of paper until the reading on the wet-bulb thermometer stops changing. Read the temperature on each thermometer.
   a. What is the temperature on the dry-bulb thermometer?
   b. What is the temperature on the wet-bulb thermometer?
   c. What is the difference in the two temperature readings?
4. Use Table 24.1 to find the relative humidity based on your temperature readings in Step 3. Look at the left-hand column labeled "Dry-Bulb Temperature," First, find the temperature you recorded in Step 3(a). Follow along to the right in the table until you come to the number that is directly below the "Difference in Temperature" (top row of the table) that you recorded in Step 3(c). This number, expressed as a percentage, is the relative humidity. What is the relative humidity at your laboratory station?

Table 24.1 Relative Humidity Table

	Difference in Temperature (°C)									
	1.0	2.0	3.0	4.0	5.0	6.0	7.0	8.0	9.0	10.0
10	88	77	66	55	44	34	24	15	6	—
11	89	78	67	56	46	36	27	18	9	—
12	89	78	68	58	48	39	29	21	12	—
13	89	79	69	59	50	41	32	23	15	7
14	90	79	70	60	51	42	34	26	18	10
15	90	80	71	61	53	44	36	27	20	13
16	90	81	71	63	54	46	38	30	23	15
17	90	81	72	64	55	47	40	32	25	18
18	91	82	73	65	57	49	41	34	27	20
19	91	82	74	65	58	50	43	36	29	22
20	91	83	74	66	59	51	44	37	31	24
21	91	83	75	67	60	53	46	39	32	26
22	92	83	76	68	61	54	47	40	34	28
23	92	84	76	69	62	55	48	42	36	30
24	92	84	77	69	62	56	49	43	37	31
25	92	84	77	70	63	57	50	44	39	33
26	92	85	78	71	64	58	51	46	40	34
27	92	85	78	71	65	58	52	47	41	36
28	93	85	78	72	65	59	53	48	42	37
29	93	86	79	72	66	60	54	49	43	38
30	93	86	79	73	67	61	55	50	44	39
31	93	86	80	73	67	61	56	51	45	40
32	93	86	80	74	68	62	57	51	46	41
33	93	87	80	74	68	63	57	52	47	42
34	93	87	81	75	69	63	58	53	48	43
35	94	87	81	75	69	64	59	54	49	44
36	94	87	81	75	70	64	59	54	50	45
37	94	87	82	76	70	65	60	55	51	46
38	94	88	82	76	71	66	60	56	51	47
39	94	88	82	77	71	66	61	57	52	48
40	94	88	82	77	72	67	62	57	53	48

Dry-Bulb Temperature (°C)

Analysis and Conclusions

1. Based on the relative humidity you found, can the air in your classroom hold more evaporated water?
2. If you wet the back of your hand, would the water evaporate and cool your skin?

Extensions

1. Suppose you exercise in a room in which the relative humidity is 100%.
   a. Would the moisture on your skin from perspiration evaporate easily?
   b. Would you be able to cool off readily? Explain.
2. Suppose you have just stepped out of a swimming pool. The relative humidity is low, about 30%. How would you feel—warm or cool? Explain.

Laboratory Notes

M O D E R N   E A R T H   S C I E N C E

Chapter 24: Water in the Atmosphere
In-Depth Investigation: Relative Humidity

Objective
In this investigation, you will use wet-bulb and dry-bulb thermometer readings to determine relative humidity.

Prelab Preparation
1. Study Graph 24.1.
   a. How many grams of water can a cubic meter of air hold at a temperature of 20°C?

   _____

   b. How many grams can the air hold at 40°C?

   _____

Graph 24.1

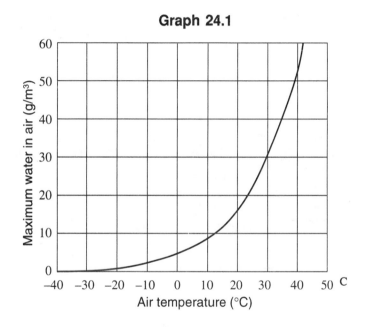

2. If a cubic meter of air at 40°C holds 40 g of water, what is the relative humidity?

   _____

Observations
1. **a.** Will the two thermometers have the same reading?

   _____

   b. If not, which thermometer will have the lower reading?

   _____

2. a. What is the temperature on the dry-bulb thermometer?

_____

  b. What is the temperature on the wet-bulb thermometer?

_____

  c. What is the difference in the two temperature readings?

_____

3. What is the relative humidity at your laboratory station? Use Table 24.1 on page 111.

_____

Analysis and Conclusions

1. Based on the relative humidity you found, can the air in your classroom hold more evaporated water?

_____

2. If you wet the back of your hand, would the water evaporate and cool your skin?

_____

Extensions

1. Suppose you exercise in a room in which the relative humidity is 100%.
  a. Would the moisture on your skin from perspiration evaporate easily?

_____

  b. Would you be able to cool off readily? Explain.

_____

_____

2. Suppose you have just stepped out of a swimming pool. The relative humidity is 30%. How would you feel—warm or cool? Explain.

_____

_____

_____

_____

_____

M O D E R N   E A R T H   S C I E N C E

Chapter 26: Climate

In-Depth Investigation: Factors That Affect Climate

Objective
In this investigation, you will explore how the angle of the sun's rays and the distribution of land and water affect climate.

Skills
analyzing, comparing and contrasting, inferring, interpreting data, predicting

Introduction
The climate of a region is the summary of the region's weather conditions and changes throughout the year. The importance of climate as a geographical factor cannot be underestimated. In additon to its affects on human life, climate determines the type of soil and vegetation found in a given area. That can then determine how the land will be used to support the animal population. Although science and engineering projects are being designed to accommodate the weather, the distribution of world populations is still primarily determined by favorable natural climates.

Materials
masking tape	graph paper	Celsius thermometers (2)
black construction paper	heat lamp	soil/water
two containers	meter stick	flashlight

Prelab Preparation
Review Chapter 26, Section 26.1 Factors That Affect Climate, pages 523–528.

Procedure
1. Roll the construction paper into a tube so that it is approximately the same size as the glass of the flashlight. Tape the edges of the paper together so that it does not unroll.
2. Attach the tube to the flashlight with the tape, as shown in Figure 26.1 on page 122. Turn on the flashlight and check to see that no light is coming from the seams you taped together.
3. Place the graph paper on a flat surface and hold the flashlight so that the edge of the paper tube is 15 cm from the graph paper and pointing straight down. Assume that the light is a sun ray and that the graph paper is the ground.
4. Draw a line around the lighted area on the graph paper. Count and record the number of squares that are completely illuminated by the light. You may need to darken the room.
5. Keep the flashlight 15 cm from another piece of graph paper, but hold the flashlight at a 45° angle to the graph paper. Repeat Step 4. Record the number of squares that are completely illuminated by the light.
6. Fill one of the containers with soil and the other with tap water. Place both containers on a flat surface next to each other.
7. Place the thermometer in the water and record the temperature.
8. Place the second thermometer in the container of soil, as shown in Figure 26.2 on page 122. The bulbs of both thermometers should be placed so that they are covered by no more than 0.5 cm of water or soil.
9. Place the heat lamp 25 cm above both containers. Record the temperature of each sample at 1, 3, 5, and 10 minute intervals after the lamp is turned on.

10. Disconnect the lamp and record the temperature of the soil and water after 5 minutes. **CAUTION: Be sure to let the heat lamp cool before storing it.**

Analysis and Conclusions

1. Compare the area of the lighted squares obtained from the two trials in Steps 4 and 5. Which area is greater?

2. What area(s) of the earth is represented by the light striking the graph paper at a 90° angle, as in Step 4? What area(s) of the earth is represented by the 45° angle, as in Step 5.

3. Why do the sun's rays from overhead cause more warming than low-angle rays?

4. What conclusion can you draw about the angle of the sun's rays during the different seasons in the United States?

5. What substance, water or soil, absorbed more heat in Step 9?

6. What substance, water or soil, lost heat faster when the heat source was turned off the Step 10?

7. What conclusion can you draw about how land and water on the earth are heated by the sun?

8. Explain how the light striking the graph paper in Steps 4 and 5 is related to the effect of heat absorption in water and soil on the earth, as simulated in Steps 9 and 10.

Extensions

1. Continue this investigation using the flashlight and graph paper. Hold the flashlight at various angles to the paper, such as at a 20° angle, 0° angle, and so on. What areas on the earth are represented on the paper at the different angles you use?

2. Considering the earth's varied climates, what conclusions can you draw concerning the angle of the sun's rays on different locations on the earth?

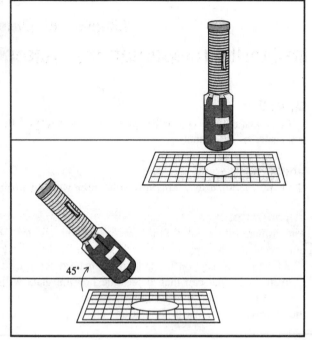

Figure 26.1

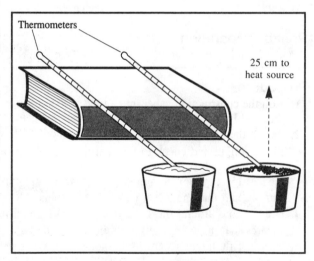

Figure 26.2

M O D E R N   E A R T H   S C I E N C E

Chapter 26: Climate

In-Depth Investigation: Factors That Affect Climate

Objective

In this investigation, you will explore how the angle of the sun's rays and the distribution of land and water affect climate.

Observations

1. How many squares on the graph paper were lit in Step 4? How many were lit in Step 5?

   _____

2. What area(s) of the earth is represented by the light striking the graph paper at a 90° angle? What area(s) of the earth is represented by the 45° angle?

   _____

3. What was the recorded temperature of the water after 5 minutes? After 10 minutes? Five minutes after the light was turned off?

   _____

   _____

Analysis and Conclusions

1. Compare the area of the lighted squares obtained from the two trials. Which area is greater?

   _____

   _____

2. What area(s) of the earth is represented by the flashlight striking the graph paper at a 90° angle? What area(s) of the earth is represented by the 45° angle?

   _____

   _____

3. Why do the sun's rays from overhead cause more warming than low-angle rays?

   _____

   _____

4. What conclusion can you draw about the angle of the sun's rays during the different seasons in the United States?

   _____

   _____

   _____

   _____

   _____

5. Which substance, water or soil, absorbed more heat in Step 9?

   _____

6. Which substance, water or soil, lost heat faster when the heat source was turned off in Step 10?

   _____

7. What conclusion can you draw about how land and water on the earth are heated by the sun?

_____

_____

8. Explain how the light striking the graph paper in Steps 4 and 5 is related to the effect of heat absorption in water and soil on the earth, as simulated in Steps 9 and 10.

_____

_____

_____

_____

Extensions

1. Continue this investigation using the flashlight and graph paper. Hold the flashlight at various angles to the paper such as at a 20° angle, 0° angle, and so on. What areas of the earth are represented on the paper at the different angles you use?

_____

_____

2. Considering the earth's varied climates, what conclusions can you draw concerning the angle of the sun's rays on different locations on the earth?

_____

_____

_____

_____

M O D E R N   E A R T H   S C I E N C E

Chapter 27: Stars and Galaxies
In-Depth Investigation: Star Magnitudes

Objective
In this investigation, you will determine the effect of distance on brightness and the relation between temperature and color.

Skills
observing, measuring, calculating, predicting

Introduction
Astronomers study the brightness, or magnitude, of stars. Except for the sun, stars are very faint and visible only at night. Thus, their brightness must be measured with a device called a photometer. An astronomical photometer consists of a surface that is sensitive to light and a device that measures the amount of light that reaches the surface. The brightest part of a star's spectrum can occur at any wavelength of optical light. Some stars look very bluish while others look redish. Color is an indication of the star's surface temperature. There is also a relation between the color of a star and its brightness. Photometers can be used to compare the colors of different light sources.

Materials
2 3-volt flashlight bulbs	masking tape	large rubber band
3 AA batteries	2 paraffin bricks	ruler
2 plastic-coated wires with	(12 × 6 cm)	desk lamp with
stripped ends (15 cm and 20 cm)	aluminum foil (12 × 12 cm)	incandescent bulb

Prelab Preparation

1. Review Chapter 27, Section 27.1 Characteristics of Stars, pages 547–548 and 550–553.
2. Review the safety guidelines for glassware safety.
3. Constructing a flashlight:
   a. Arrange the bulbs and batteries as shown in Figure 27.1. Using masking tape, attach the wires to the batteries and bulbs. The bulb should be on. If it is not, study the wiring diagram again and make adjustments.

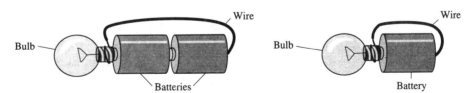

Figure 27.1

   b. Tape the flashlight arrangement together so it can be moved. Be sure to leave the wire loose at the end of the battery so that you can turn your flashlight on and off.
4. Constructing a photometer: Fold the aluminum foil in half with the shiny side facing out, and place it between the paraffin bricks. Hold the pieces together with a rubber band (see Figure 27.2).

Procedure

1. Place the two flashlights on a table about 2 m apart. Place the photometer between them with the largest sides of the bricks facing each flashlight bulb, as shown in Figure 27.2.

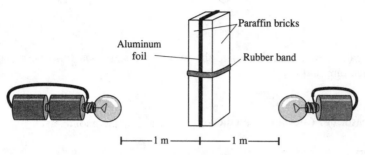

Aluminum foil   Paraffin bricks

Rubber band

├────1 m────┼────1 m────┤

Figure 27.2

2. Turn on both flashlights, and turn off all room lights.
3. Move the paraffin photometer until both sides are equally bright. Measure the distance, in centimeters, from each flashlight bulb to the center of the photometer. Record these measurements in your lab report. Which way did you move the photometer to make it record equal brightness on both sides?
4. Square the distances you recorded in Step 3. Record these distances in your lab report.
5. Incandescent light bulbs have filaments inside that glow at a temperature much cooler than the sun's surface. Place the photometer between the desk lamp and a window on a bright day. Sunlight coming through a window will be the same color as the sunlight outdoors. Turn off any fluorescent ceiling lighting, and turn on the lamp.
6. Compare the color differences between the paraffin sides of your photometer. Which light, the light bulb or sunlight, is more yellow? Which light is more white?
7. Darken the room once again, and compare the colors of the bulb powered by one battery with the colors of the bulb powered by two batteries.

Analysis and Conclusions

1. As you move away from a light source, the brightness decreases in relation to the square of the distance. The ratio of the square of the distances you calculated in Step 4 is equal to the ratio of the brightnesses of the bulbs. What is the ratio of the square of the distance of the two-battery flashlight to that of the one-battery flashlight? What does this tell you about the relation of the brightnesses of the two flashlights?
2. An astronomer knows that two stars have the same spectra and should be the same brightness. Yet one is four times fainter than the other. How much farther away is the fainter star?
3. Based on the results of the investigation, would you expect a white star or a yellow star to be hotter?
4. Using your knowledge of the spectrum, would you expect a white star to be hotter or cooler than an orange star? Predict whether a blue star is hotter or cooler than a white star. Also predict whether a red star is hotter or cooler than an orange star.

Extensions

1. Find an incandescent bulb controlled by a dimmer. Watch the color of the light as it fades. Does it get more yellow or more white? Explain why.
2. If a red star is very cool with a low absolute magnitude, why are bright red stars visible in the sky?

M O D E R N   E A R T H   S C I E N C E

Chapter 27: Stars and Galaxies
In-Depth Investigation: Star Magnitudes

Objective
In this investigation, you will determine the effect of distance on brightness and the relation between temperature and color.

Observations
1. What are the distances from each bulb to the center of the photometer?

   _____

   _____

2. Which way did you move the photometer to make it record equal brightnesses on both sides?

   _____

3. Square the distances you recorded in Step 1 of the Observations.

   _____

4. Compare the color differences between the paraffin sides of your photometer.
   a. Is the light bulb or sunlight more yellow?

   _____

   b. Is the light bulb or sunlight more white?

   _____

5. Compare the colors of the bulb powered by a single battery and the bulb powered by two batteries in the darkened room. What did you observe?

   _____

   _____

Analysis and Conclusions
1. As you move away from a light source, the brightness decreases in relation to the square of the distance. The ratio of the square of the distances you calculated in Step 3 of the Observations is equal to the ratio of the brightnesses of the bulbs. What is the ratio of the square of the distance of the two-battery flashlight to that of the one-battery flashlight? What does this tell you about the relation of the brightnesses of the two flashlights?

   _____

   _____

   _____

2. An astronomer knows that two stars have the same spectra and should be of the same brightness. Yet one is four times fainter than the other. How much further away is the fainter star?

   _____

   _____

3. Based on the results of the investigation, would you expect a white star or a yellow star to be hotter?

_____

4. Would you expect a white star to be hotter or cooler than an orange star? Predict whether a blue star is hotter or cooler than a white star. Predict whether a red star is hotter or cooler than an orange star.

_____

_____

_____

Extensions

1. Find an incandescent bulb controlled by a dimmer. Watch the color of the light as it fades. Does it get more yellow or more white? Explain why.

_____

_____

2. If a red star is very cool with a low absolute magnitude, why are bright red stars visible in the sky?

_____

_____

_____

M O D E R N   E A R T H   S C I E N C E

Chapter 28: The Sun

In-Depth Investigation: Size and Energy of the Sun

Objective

In this investigation, you will perform a simple experiment from which you can calculate the sun's diameter. Then you will collect energy from sunlight and estimate the amount of energy produced by the sun.

Skills

observing, measuring, calculating, inferring, predicting

Part I Introduction

The sun is an average of 150 million kilometers away from the earth. Yet scientists have been able to measure the sun's size and distance from the earth. Scientists use complicated astronomical instruments to make these measurements. However, it is possible to measure the sun's size with a simple instrument called a solar viewer and the knowledge of the relation between the size of an object and the size of the object's image as seen through the viewer.

Materials

shoe box with lid	piece of very thin	glass jar with lid
aluminum foil	sheet metal (2 × 8 cm)	modeling clay
safety pin	black paint, flat finish	Celsius thermometer
index card	scissors	masking tape
metric ruler	desk lamp with 100-watt bulb	pencil

Prelab Preparation

1. Review Chapter 28, Section 28.1 Structure of the Sun, pages 571–573.
2. Review the guidelines for glassware safety.
3. Describe how a pinhole image of the sun is formed.
4. Constructing a pinhole solar viewer:
    a. Refer to Figure 28.1 when making your solar viewer. Cut a hole with a diameter of about 3 cm in the center of one end of the shoe box. Tape the index card to the inside wall of the box opposite the hole. Tape the foil over the hole and use the pin to make a tiny hole in the center of the foil.

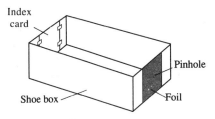

Figure 28.1

5. Constructing a solar collector:
    a. Punch a hole in the jar lid or use one already prepared by your teacher. **CAUTION: When using holepunching equipment, you must be supervised by an adult.**

    b. Gently bend the sheet-metal piece around a pencil to shape it. Then gently place the metal piece around the thermometer bulb so it fits snugly. Be careful not to press too hard. Next bend the remaining metal outward, as shown in Figure 28.2, to collect as much sunlight as possible.

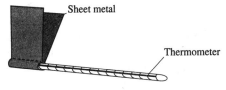

Figure 28.2

    c. Paint the sheet metal black.

d. Slip the top of the thermometer through the hole in the lid. Mold the clay around the thermometer on the top and bottom of the lid to hold the thermometer steady. Then secure the thermometer and clay to the lid with masking tape. Place the lid on the jar so that the thermometer bulb is in the middle of the jar, as shown in Figure 28.3.

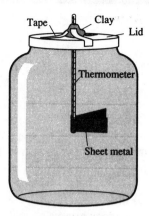

Figure 28.3

Procedure

1. Aim the pinhole in the solar viewer toward the sun. **CAUTION: You should never look directly at the sun.** Adjust the box until the sunlight forms a small image on the index card at the other end of the viewer.
2. Carefully measure and record the diameter of the image. Measure and record the distance between the image and the pinhole.
3. The ratio of the diameter of the sun's image to its distance to the pinhole is equal to the ratio of the sun's diameter divided by the distance to the sun, as shown by the following equation:

$$\frac{\text{sun's diameter}}{\text{sun's distance to pinhole}} = \frac{\text{image diameter}}{\text{image's distance to the pinhole}}$$

If the distance between the sun and the earth is 150 million kilometers, calculate the diameter of the sun using your data.

Analysis and Conclusions

1. What is the sun's diameter as given in the textbook?
2. Explain any difference in your answer and the textbook value.

Part II Introduction

Energy, whether from a light bulb or from the sun, spreads out in all directions. The amount of energy received decreases with the square of the distance between the energy source and the point that receives the energy. In other words, if a light is moved three times farther away from its original distance from a point, the energy received at that point is nine times less.

Procedure

1. Place the solar collector so that the black metal is angled toward the sun. If it is cool outside, use sunlight shining through an open or untinted window. **CAUTION: You should never look directly at the sun.**
2. Watch the temperature reading on the thermometer until it reaches a maximum value. Record this value.
3. Allow the collector to cool in the shade until it reaches room temperature.
4. Place the lamp at the end of a table. Remove any reflector or shade from the lamp.
5. Turn the collector toward the lamp and put the collector about 30 cm from the lamp.
6. Turn on the lamp and wait one minute. Gradually move the collector toward the lamp. Watch the temperature carefully. Stop when the temperature reaches the maximum that was achieved in sunlight. If the temperature rises above that level, move the collector back and let it cool to room temperature. Repeat the experiment and watch the temperature's rise more carefully.
7. Once the temperature has stabilized at the same level reached in sunlight, record the distance between the center of the lamp and the thermometer bulb.

Analysis and Conclusions

1. The collector absorbed as much energy from the sun at a distance of 150 million km as it did from the 100-W bulb at the distance you measured. Therefore, the following equation can be used.

$$\frac{\text{power of the sun (in watts)}}{(\text{distance to sun})^2} = \frac{\text{power of the lamp (in watts)}}{(\text{distance to lamp})^2}$$

The distance to the sun is 1.5×10^{13} centimeters. Use the equation above to calculate the power of the sun. Be sure that you express both distances in the same units.

2. The sun's power is quoted as 3.7×10^{26} watts. Compare your answer with this value. Describe any sources of measuring inaccuracy with your method. Do not be surprised if your answer is 100 times too large or small. This is a very crude method to approximate a very large number.

Extension

Would the experiment have worked with a flourescent bulb? Explain.

Laboratory Notes

M O D E R N   E A R T H   S C I E N C E

Chapter 28: The Sun

In-Depth Investigation: Size and Energy of the Sun

Objective

In this investigation, you will perform a simple experiment from which you can calculate the sun's diameter. Then you will collect energy from sunlight and estimate the amount of energy produced by the sun.

Prelab Preparation

Describe how a pinhole image of the sun is formed.

_____

_____

Part I Observations

1. What is the diameter of the sun's image?

_____

2. What is the distance between the image and the pinhole?

_____

3. The ratio of the diameter of the sun's image to its distance to the pinhole is equal to the ratio of the sun's diameter divided by the distance to the sun as shown by the equation:

$$\frac{\text{sun's diameter}}{\text{sun's distance to pinhole}} = \frac{\text{image diameter}}{\text{image's distance to the pinhole}}$$

If the distance between the sun and the earth is 150 million kilometers, calculate the diameter of the sun using your data.

_____

Analysis and Conclusions

1. What is the sun's diameter as given in the textbook?

_____

2. Explain any difference in your answer and the textbook value.

_____

_____

_____

Part II Observations

1. What is the temperature reading on the thermometer when the solar collector faces the sun?

_____

2. What is the distance between the center of the lamp and the thermometer bulb?

_____

133

Analysis and Conclusions

1. The collector absorbed as much energy from the sun at a distance of 150 million km as it did from the 100-W bulb at the distance you measured. Therefore, the following equation can be used.

$$\frac{\text{power of the sun (in watts)}}{(\text{distance to sun})^2} = \frac{\text{power of the lamp (in watts)}}{(\text{distance to lamp})^2}$$

The distance to the sun is 1.5×10^{13} centimeters. Use the equation above to calculate the power of the sun. Be sure that you express both distances in the same units.

_____

2. The sun's power is quoted as 3.7×10^{26} watts. Compare your answer with this value. Describe any sources of measuring inaccuracy with your method. Do not be surprised if your answer is 100 times too large or small. This is a very crude method to approximate a very large number.

_____

_____

_____

Extension

Would the experiment have worked with a flourescent bulb? Explain.

_____

_____

_____

M O D E R N   E A R T H   S C I E N C E

Chapter 29: The Solar System
In-Depth Investigation: Crater Analysis

Objective
In this investigation, you will experiment with making craters to discover the effect of speed and projectile angle on the craters formed.

Skills
measuring, observing, inferring, applying, predicting

Introduction
All the inner planets—Mercury, Venus, Earth, and Mars—have many features in common. They are all mostly solid rock with metallic cores; they have no rings and from zero to two moons each; and every inner planet has bowl-shaped depressions called *impact craters*. Impact craters are formed as a result of collisions between the planets and rocky objects traveling through space. Most of these collisions took place during the formation of the solar system. Mercury's entire surface is covered with these craters, while very few still are evident on the surface of the earth. Many of the moons of the inner and outer planets also are heavily cratered.

Materials
meter stick	water	scissors	marker
plaster of Paris	shoe box	6 toothpicks	safety goggles
marbles, 1 of them large	protractor	masking tape	lab apron

Prelab Preparation
1. Review Chapter 29, Section 29.2 The Inner Planets, page 594.
2. Review the guidelines for eye safety.
3. Place the top of a toothpick in the center of a 6-cm-long piece of masking tape. Fold the tape in half around the toothpick so that they form a small flag and flagpole. On the flag, write the letter *A*. Repeat this procedure with each of the other toothpicks, labeling them *B* through *F*.

Procedure
1. Mix plaster of Paris with water, according to instructions. Spread your mixture in the bottom of the shoe box. Make your surface about 4 cm thick. The surface should be as smooth as possible.
2. Allow your surface to dry until it is no longer soupy, but not yet rigid.
3. **Put on your safety goggles.** Drop a marble onto the plaster from a height of 50 cm above the surface. Quickly remove the marble, without damaging the crater if possible. Place flag *A* next to the crater to label it crater *A*.
4. Repeat Step 3 with a marble dropped from a height of 1 m and another dropped from a height of 25 cm. Use the flags to mark craters *B* (1-m drop) and *C* (25-cm drop).
5. Repeat Step 3 using the large marble dropped from a height of 1 m. Label the crater formed *D*.
6. Using your protractor to measure the angle as shown in Figure 29.1 on page 136, have your partner tilt the box at a 30-degree angle to the table. Be sure to hold the box steady. Then drop a marble vertically from a height of 50 cm. Label the crater *E*.

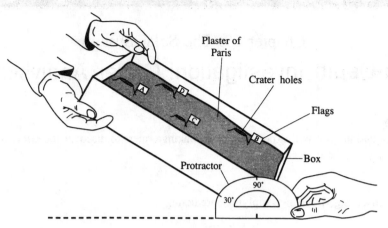

Plaster of
Paris

Crater holes

Flags

Box

Protractor

90°
30°

Figure 29.1

7. Repeat Step 5 using an angle of 45 degrees. Label the crater *F*.
8. Allow the plaster of Paris to harden. Write a description of each crater and the surrounding area.

Analysis and Conclusions

1. Which crater was formed by the marble with the highest velocity? What is the effect of velocity on the characteristics of the crater formed?
2. Study the shapes of craters *A*, *E*, and *F*. How does the angle of incidence affect the shape of the craters formed?
3. Compare craters *A*, *B*, and *D*. How do they differ? What caused this difference? Is the difference in the masses of the objects a factor? Explain.

Extension

Which type of crater formation have we not studied? Devise a method of studying this cratering process using the model you made.

M O D E R N   E A R T H   S C I E N C E

Chapter 29: The Solar System
In-Depth Investigation: Crater Analysis

Objective
In this investigation, you will experiment with making craters to discover the effect of speed and projectile angle on the crater formed.

Observations
Write a description of each of the craters and the surrounding area.

Crater A _____

_____

_____

Crater B _____

_____

_____

Crater C _____

_____

_____

Crater D _____

_____

_____

Crater E _____

_____

_____

Crater F _____

_____

_____

Analysis and Conclusions
1. Which crater was formed by the marble with the highest velocity? What is the effect of velocity on the characteristics of the crater formed?

_____

_____

2. Study the shapes of craters *A*, *E*, and *F*. How does the angle at which the marble strikes the surface affect the shape of the craters formed?

_____

_____

_____

_____

3. Compare craters *A*, *B*, and *D*. How do they differ? What caused this difference? Is the difference in the masses of the objects a factor? Explain.

_____

_____

_____

_____

_____

_____

_____

Extension

Which type of crater formation have we not studied? Devise a method of studying this cratering process using the model you made.

_____

_____

_____

_____

M O D E R N   E A R T H   S C I E N C E

Chapter 30: Moons and Rings
In-Depth Investigation: Galilean Moons of Jupiter

Objective
In this investigation, you will verify that the orbital motions of Jupiter's moons obey Kepler's third law.

Skills
measuring, recording, applying a theoretical model, predicting

Introduction
A German astronomer named Johannes Kepler developed three laws that explained most aspects of planetary motion. The third law—the law of periods—explained the relation between a planet's distance from the sun and the planet's period of orbit. The period of orbit is the time required for the planet to make a complete revolution around the sun. According to the law of periods, the cube of the average distance of the planet from the sun is proportional to the square of the planet's period. It can be expressed mathematically as

$$K \times r^3 = p^2$$

where r is the distance from the sun, p is the period, and K is a constant. Kepler's third law also may be applied to moon's orbiting a planet, where r is the distance of a moon to the planet and p is the moon's period, or the time required for the moon to make one revolution around the planet.

Materials
metric ruler
calculator

Prelab Preparation
1. Review Chapter 30, Section 30.4 Satellites of Other Planets, pages 631–632. Also review Chapter 29, Section 29.1 Models of the Solar System, page 593.
2. Find out what is meant by a constant and a variable.

Procedure

1. Two telescope eyepiece views in Figure 30.1 show how Jupiter and its four largest, or Galilean, moons appear through a telescope on the earth at midnight on the 9th and 19th of a month. Compare these illustrations with Figure 30.2 on page 140, which shows the path of each moon as it orbits Jupiter during the same month. The central

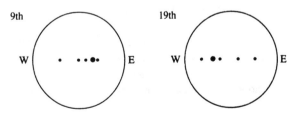

Figure 30.1

horizontal band on the chart represents Jupiter. When a moon's path crosses in front of this band, the moon is in front of the planet. When a moon's path crosses behind this band, the moon is behind Jupiter.

a. List the days when each of Jupiter's moons crosses in front of the planet.

b. List the days when each of the moons is behind Jupiter.

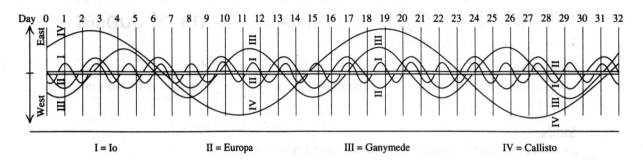

I = Io   II = Europa   III = Ganymede   IV = Callisto

Figure 30.2

2. Use the data in Table 30.1 to test Kepler's third law. Calculate p^2 and r^3 for each of the planets. Record your results in the table. Then calculate K for each planet using Kepler's third law, $K = p^2/r^3$. Record your results in the table. Is K a constant?

Table 30.1 The Planets

Planet	r (in billions of kilometers)	p (in Earth years)	r^3	p^2	K
Mercury	0.058	0.24			
Venus	0.108	0.62			
Earth	0.150	1			
Mars	0.228	1.88			
Jupiter	0.778	11.86			
Saturn	1.427	29.46			
Uranus	2.869	84.01			
Neptune	4.486	164.8			
Pluto	5.890	247.7			

3. Draw Jupiter and its moons as they would appear from the earth at midnight on the 2nd and 26th of the month.

4. Draw Jupiter's moons on the first day of the month that all four moons are on the same side of the planet. Give the date.

5. Give a date when only two moons will be visible. Name the two visible moons.

6. Follow each moon's motion on the chart. Find the length of time, in earth days, required for each moon to orbit Jupiter. To do this, measure the time between two points when the moon is in exactly the same position on the same side of Jupiter. Record your answers in Table 30.2 in your laboratory report.

Table 30.2 Galilean Moons

Moon	p (in Earth days)	Scale r (in cm)	p^2	r^3	K
Io					
Europa					
Ganymede					
Callisto					

7. Measure the scale distance between the maximum outward swing of each moon and the center of Jupiter in centimeters. Record your answers in Table 30.2.

8. Square each period measurement and record the answer in Table 30.2. Cube each distance measurement and record the answer in the table.

9. Use your results to test Kepler's third law. Because $K = p^2/r^3$, divide p^2 by r^3 for each moon to find K. Record your results in Table 30.2.

Analysis and Conclusions

1. Will you see all four of Jupiter's largest moons each time you look at Jupiter through a telescope or binoculars? Explain.

2. Jupiter's moons look like dots in a telescope. You cannot tell them apart by their appearance. If you had no charts, how could you identify each moon?

3. After you solve for K for each moon, study your results. Is K a constant?

Laboratory Notes

M O D E R N   E A R T H   S C I E N C E

Chapter 30: Moons and Rings

In-Depth Investigation: Galilean Moons of Jupiter

Objective

In this investigation, you will verify that the orbital motion of Jupiter's moons obey Kepler's third law.

Prelab Preparation

Find out what is meant by a constant and a variable.

_____

_____

Procedure

1. a. Using Figure 30.2 on page 140, list the days when each of Jupiter's moons cross in front of the planet.

_____

_____

b. List the days when the moon's are behind the planet.

_____

_____

2. Use the data in Table 30.1 to test Kepler's third law. Calculate p^2 and r^3 for each of the planets. Record your results in the table. Then calculate K for each planet using Kepler's third law, $K = p^2/r^3$. Record your results in the table.

Table 30.1 The Planets

Planet	r (in billions of kilometers)	p (in Earth years)	r^3	p^2	K
Mercury	0.058	0.24			
Venus	0.108	0.62			
Earth	0.150	1			
Mars	0.228	1.88			
Jupiter	0.778	11.86			
Saturn	1.427	29.46			
Uranus	2.869	84.01			
Neptune	4.486	164.8			
Pluto	5.890	247.7			

3. According to your results in Table 30.1, is K a constant?

_____

4. Draw Jupiter and its moons as they would appear from the earth at midnight on the 2nd and 26th of the month in the eyepiece of the telescope shown in Figure 30.3 on page 144.

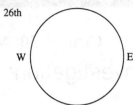

Figure 30.3

5. In Figure 30.4, draw Jupiter's moons on the first day of the month that all four moons are on the same side of the planet. Give the date.
6. Give a date when only two moons will be visible. Name the two visible moons.

_____

_____

_____

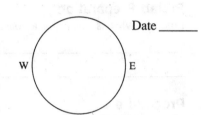

Date _____

Figure 30.4

7. In Table 30.2, record the length of time in earth days required by each moon to orbit Jupiter.

Table 30.2 Galilean Moons

Moon	p (in Earth days)	Scale r (in cm)	p^2	r^3	K
Io					
Europa					
Ganymede					
Callisto					

8. In Table 30.2, record the scale distance between the maximum outward swing of each moon and the center of Jupiter in centimeters.
9. Square each period measurement and record the answer in Table 30.2. Cube each distance measurement and record the answer in the table.
10. Use your results to test Kepler's third law. Because $K = p^2/r^3$, divide p^2 by r^3 for each moon to find K. Record your results in Table 30.2.

Analysis and Conclusions

1. Will you see all four of Jupiter's largest moons each time you look at Jupiter through a telescope or binoculars? Explain.

_____

2. Jupiter's moons look like dots in a telescope. You cannot tell them apart by their appearance. If you had no charts, how could you identify each moon?

_____

3. After you solve for K for each of the moons, study your results. Is K a constant?

_____